Wargaming in History

Series Editor
Stuart Asquith

The Second Anglo-Boer War

·····

Edwin Herbert

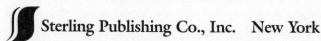
Sterling Publishing Co., Inc.   New York

Published in 1990 by Sterling Publishing Company, Inc.
387 Park Avenue South, New York, N.Y. 10016

Distributed in Canada by Sterling Publishing
c/o Canadian Manda Group, P.O. Box 920, Station U
Toronto, Ontario, Canada M8Z 5P9

© 1990 by Argus Books

ISBN 0-8069-7327-7

First published in the U.K. in 1990 by Argus Books.
Published by arrangement with Argus Books.
This edition for sale in the United States, Canada,
and the Philippine Islands only.

Library of Congress Cataloging-in-Publication Data

Herbert, Edwin
The second Anglo-Boer War/Edwin Herbert
p.   cm.--(Wargaming in history)
ISBN 0-8069-7327-7
1. War games--Design and construction.   2. South African
War 1899–1902.
I. Title.   II. Series
U310. H425 1989
793.9′2--dc20   69-49078
                                                                          CIP

Typeset by Photoprint, Torquay, Devon.
Printed and bound by LR Printing Services Ltd, Edward Way,
Burgess Hill, West Sussex, RH15 9UA, England.

CONTENTS

INTRODUCTION

The Second Anglo-Boer War had its origins in the First Boer, or Transvaal, War of 1880–81. The Boer resentment of Imperial influence culminated in a campaign in which British forces were defeated at Bronkhurst Spruit, Laing's Nek and Majuba Hill. The Conservative administration under Gladstone sought terms and granted the Transvaal self-government, with Britain retaining control of foreign affairs. But neither side was really satisfied with the peace. The British military authorities felt that they had been humiliated without redress; and the Boers had developed sufficient confidence in their fighting prowess to feel that they should demand complete independence. Matters were brought to a head in 1896 when the Jameson Raid attempted to foster revolt among the Uitlanders in Johannesburg. After that, war was just a question of time.

War broke out between Great Britain and the South African Republic of the Transvaal on Wednesday 11 October 1899. The Orange Free State immediately honoured its treaty obligations to support the Transvaal. The war was known to the Afrikaners as the Second War of Freedom (Tweede Vryheidsoorlog). In Britain the popular view was that it would all be over within a few months, probably by Christmas. In fact, it was to last until 31 May

Dr Jameson and the 'African' Mounted Infantry by Britains.

1902 and to involve 550,000 combatants and many civilians, with the loss of over 70,000 lives, most of them by disease. It was the largest campaign ever fought by British forces in Africa and, in terms of cost and numbers involved, it surpassed all other Victorian campaigns, including the Crimean War.

In many ways the fascination in studying the war lies in its human interest rather than in its military value, although it provided some sharp lessons for the British Army. The present political position in South Africa, including the emergence of the National Party, has many of its roots in the war.

The war falls into three distinct phases: the opening Boer offensives against Cape Colony and Natal; the counter-strikes by Roberts and the capture of the republican capitals of Bloemfontein and Pretoria; and the final guerilla campaigns. It is not the purpose of this book to provide a definite narrative of the events that constituted these three phases: this has been done in other works (see, for example, *The Boer War* by Thomas Pakenham). Rather, this is a book on how to wargame the special aspects of the war. It illustrates the three phases of the war by reference to the set-piece Battle of Elandslaagte in October 1899, the strategic entrapment of the Boers at Paardeberg in February 1900 and a skirmish action between the Bushveldt Carabineers and Boer guerillas at Duivelskloof in August 1901. To set these actions in context, a brief diary of events is given in Appendix A.

In addition, the book deals with the organisation, uniforms and weapons of the Boers and the British with their respective battle tactics, morale and casualties. Its aim is to provide the necessary background information for those writing and using wargame rules for the period; and some suggestions on rules are put forward. But why a book on wargaming the war? Is it not a little strange to throw dice to simulate a real war? Well, Churchill was a wargamer in his youth and a combatant with both the pen and the sword in his early career; and he said that his survival to a ripe old age depended on more than one occasion on throwing a double six!

If you can model the war in terms of developing a good set of wargame rules, incorporating an element of luck as well as the special features of contemporary technology, you will begin to realise what this campaign, at the turn of the century, was all about. On the Boer side, individuals were prepared to die for independence. On the British side, massed troops were prepared to die for patriotism. The Boer commanders tried to minimise casualties amongst their men: white lives were precious. The

British commanders were less concerned. Indeed, as shown by their actions, some had an absolute indifference to losses. Effective wargame rules must allow for these factors.

Fighting a wargame with a Boer army is an exciting activity for those with strong nerves. You either do extremely well and hardly lose a figure, or else you get annihilated. In the national colonial competitions that have been held from time to time, Boer armies have done quite well. The only disasters that have occurred have been against hordes of Mahdist light cavalry, which is hardly a realistic encounter. A British wargames army of the period is also effective, providing a balanced combination of arms in an 1899–1900 order of battle or mobile, well-disciplined mounted infantry in a 1901–02 wargames force. But what were the respective tactics of the two sides? How was it that a small army of farmers could defy the might of the British Empire for so long?

1 ORGANISATION AND FIELD TACTICS

THE BOERS

Boer forces consisted of a muster of all available armed men, each of whom had to supply his own horse, rifle and provisions for eight days. After this period, the state assumed responsibility for feeding the commandos. Apart from the regular artillery and police, the men were unpaid. Rifles were sold at cost to burghers but given to poor burghers who could not afford to pay. The Boers had a natural eye for country and could judge distances very well in the clear atmosphere of their homeland. Their horses enabled them to rush from one position to another very quickly. They fought as irregulars and were loosely organised in commandos each led by a Commandant and consisting of units of perhaps 150–200 men under a Field-Cornet (veldkornet) in turn composed of sections of 25 men or so under a Corporal

A Boer Vedette.

(korporaal). The approximate equivalents in the British Army would be Colonel, Major and Lieutenant respectively. In the guerilla phase of the war the sections became standardised units of 12 men.

The Head of the State Army was the Commandant-General. Under him were Assistant Commandant-Generals and various Fighting-Generals. A Krijgsraad or Council of War decided battle plans by majority decision – so that it was possible for 16 Corporals to outvote 15 senior officers. In this context it was often the personality of the general concerned that swayed the decision. He had to persuade his men, however; he could not order them.

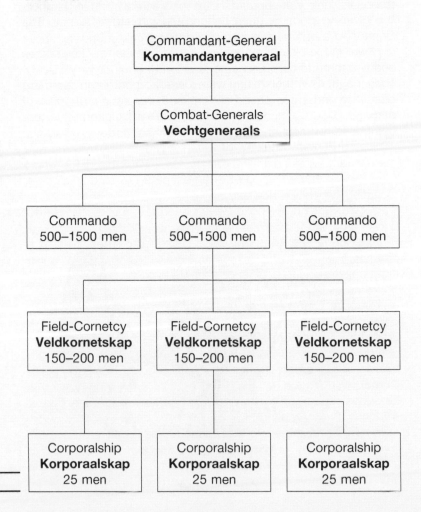

This is a convenient organisation at a scale of 25 men to a figure, as each Boer figure can be used to represent one corporalship. Remember that three figures will be required for each Boer: a mounted figure, a dismounted figure and a horse. Horseholder figures are not essential as Boer horses were generally picketed. Apart from the normal commando organisation, based on a particular district, there were special units such as the Africander Cavalry Corps (ACC), which consisted of a flying column of 150 men under Commandant Malan: this can be represented by a small unit of six figures.

In 1899 the Staatsartillerie of the Orange Free State, under Major Albrecht, consisted of five officers and 400 men, including reservists, and can therefore be represented by 16 figures. In 1901, having lost all its guns, the corps was reformed as mounted infantry.

The regular strength of the Transvaal Staatsartillerie, commanded by Commandant Trichardt, was also about 400 men, or 16 figures again. With all reservists called up, its strength was 800 men. The corps comprised mountain, siege, field and 'cavalry' batteries, together with telegraph, supply, medical and veterinary branches and a band. The other uniformed force, the ZARPs, consisted of 300 mounted officers and men, 600 officers and men on foot and 60 black auxiliaries. On a war footing, the whole force was mounted. It can be represented by, say, 36 figures, including two black auxiliary sections, at a wargames scale of 1:25. Two-thirds of the police were based at Johannesburg under G M T Van Dam; one-quarter at Pretoria; and the remainder at Fort Hendrina and Krugersdorp or in small detachments at outstations. The Swaziland Police Force consisted of 300 men under eight officers and 27 NCOs, or 12 figures: unlike the ZARPs, they operated as part of the local commando.

A popular Commandant would attract many more men to his commando than an unpopular one. Since he was only leader at the vote of his men, a Commandant could find it difficult to control his commando. Moreover, the average burgher did not take kindly to orders outside the normal course of duty. Field-Cornet A L Thring of the Kroonstadt Commando gained a notorious reputation for having a roll-call and an inspection of weapons every morning!

In contrast to the sharp distinction between officers and men in the British Army, the Boers adopted more democratic attitudes and the officers were elected by their men. This was at the same time a strength and a weakness. It was a strength in the sense that the men were volunteers fighting for their country in a band of brothers. It was a weakness in the sense that the Boers could go home when they had had enough or they could refuse to carry out orders with which they did not agree. It is likely that Mafeking could have been taken if the besieging Boer forces had followed the orders of their Commandant-General. Hillegas (p. 83) advances the paradox that all the successes of the Boer Army were the result of the fact that every burgher was a general, and to the same cause might be attributed almost every defeat. A severe weakness of the Boers was that they never exploited their successes by counter-attacks.

Commandos theoretically consisted of all the men of the district between the ages of 16 and 60. In practice old men and young boys also joined up. With the army of General Cronje captured at Paardeberg were no less than 100 burghers under 16 (it is unlikely they were all non-combatants) and among those who escaped from the laager were two Bloemfontein boys named Roux, aged 12 and 14 (Hillegas, p. 62). Grey-haired men went out on scouting duty and scaled hills with alacrity. Jan van der Westhuizen of the Middleberg Commando was 82; while Piet Kruger with the Rustenberg Commando was 86. However, it was the younger men who bore the brunt of the fighting: most of those killed or wounded were between 17 and 30.

There was no medical examination of fitness: men with only one leg, one arm or one eye rode with the commandos and men who wore spectacles were numerous – contemporary photographs do not record this: perhaps spectacle-wearers took their spectacles off in front of the camera. Several were suffering from various kinds of illness. Hillegas (p. 73) doubted whether more than 20% of Boer burghers in commandos would have been accepted in a European or the American Army.

Young or old, they were skilled fighters and their defensive tactics bloodied the British time and time again in the initial stages of the war. They would choose a good defensive position, bunch their horses behind cover, remain hidden in their trenches or rifle-pits and wait for the British to attack. The British generals, for the most part, obligingly did just that. The Boers employed three main types of fire: carefully aimed, individual fire at long distance; heavy, continuous fire directed at an attacking force at medium range; and 'snapping' fire at close quarters.

The Boers knew the terrain and would fight from their position for as long as they could with safety, causing maximum casualties among the attacking forces, and they would then withdraw to fight another day. Retreating at the proper time was one of the arts of the Boer rifleman. At Elandslaagte, the foreigners under General Kock did not gauge the proper moment for retreat and were almost annihilated because of their indiscretion.

The Boers took great care not to be outflanked or to be caught in the open and did not use the sword or bayonet, which they regarded as an inhuman weapon. The British, apart from a few actions, such as the final phase of Elandslaagte, did not exploit their advantage in mêlée. The Boers were so unaware of the use of the bayonet as a weapon that they did not relieve one group of captured Highlanders of the bayonets in their frogs. The Highlanders promptly used them to escape.

Laagers of supply wagons were still used to rally Boer forces behind the battle line. Since the Boers were in general more mobile than the Khakis, being used to travelling light, and because they invariably outscouted the British, only in a few cases were these laagers ever captured. The main exception to the rule was Paardeberg, where the Boers left their escape too late and were surrounded.

But the Boers did not always lie in wait for the attacker. At the Platrand fight near Ladysmith on 6 January 1900, the Boers charged British positions so fiercely that they drove the Khakis out and captured the positions with little loss. Deneys Reitz (p. 134) describes how his commando was so upset at seeing the British cheer when an attack from a supporting commando was driven off that the Boers charged the British position regardless of casualties and overwhelmed it. In this case there was a positive morale effect from seeing a supporting unit flee – invariably wargame rules apply a minus factor.

As the war dragged on, Boer numbers were reduced – although it tended to be the best fighting men who were left. Many Boers became prisoners of war and were sent to St Helena, Bermuda or Ceylon. At the end of the war, as many as 30,000 Boers were prisoners. Even at the beginning of the war, when the Boers had superior numbers, the loose discipline inherent in the structure of the commandos did not allow the best deployment of men. There were hundreds of Boers in the Natal laagers – the Bible Readers – who were never engaged in any battle and who never fired a shot in the first six months of the war. However, there were hundreds of eager volunteers who took part in all the actions, whether their commando was supposed to be in action or not. Overall, there was a lack of co-ordination that made it difficult to send reinforcements to the right place at the right time.

A Boer commando could move at an average speed of 5–6 miles per hour. In emergencies it could cover 7–8 miles in an hour. Scouts would ride up to six miles away on each flank, scouring the veld for signs of the Khakis. Even with wagons, for which mules rather than oxen were used, a Boer column could travel at 3–4 miles per hour, unless the train was so large that oxen had to be used and mobility was thereby impeded. Cronje's vast wagon train was reduced to a speed of ten miles a day.

The British army with its transport wagons was lucky to achieve 2 miles an hour. Whereas a Boer rider and equipment weighed about 250 lbs, a British rider with his heavier saddle probably weighed nearer 400 lbs. On a forced march, riding his two horses alternately, a Boer could cover 60–70 miles a day, far outstripping most of the Khakis (Hillegas, p. 129).

The real strength of the commandos probably never exceeded 30,000 men at any one time, of whom half were unwilling to fight very much. Not all the burghers from a particular district could be called up at the same time; there was a need to protect the homesteads and to secure the borders against raids from hostile Africans. In December 1899, when Christmas packages were distributed, there were only 26,000 men in the fleid to receive them. At the beginning of that month, there were about 13,000 armed Boers in Natal, 12,000 in the Orange Free State and its border regions and 5,000 in the Transvaal and its border regions.

Various foreign volunteers served with the Boer armies, either in the commandos or in special units. The accompanying table gives an indication of the numbers involved.

There were also a small number of Austrians under Baron von Goldeck. Each of the foreign volunteers was supplied with a horse and equipment, valued at £35, and was given better food than the burghers. They were tolerated rather than welcomed with open arms by the Boers. Because of their tendency to stay and fight, the foreigner volunteers lost more men proportionately than the local burghers. The Hollander and German contingents under Schiel suffered severe casualties at Elandslaagte and the Scandinavian Corps was virtually wiped out at Magersfontein: in the latter case only seven men out of 53 escaped. Comte Georges Villebois-Mareuil eventually led a combined foreign brigade in a single commando and became the first vechtgeneraal.

After the formal actions, the Boers split into the bittereinders or 'bitter-enders' and the hensoppers or 'hands-uppers'. It is a measure of the grim determination of the 'bitter-enders' that, at

Nationality	Commander	Own Unit	Commandos
Legion of France	Comte Georges Villebois-Mareuil	300	100
Hollander Corps	Commandant Jan Smoronberg	400	250
American Volunteers (Hassell's Scouts)	Commandant John A Hassell	150	150
Irish Brigade	Colonel John Franklin Blake	100	–
	Arthur Lynch	100	
Russian Scouts (Don Cossacks)	Commandant Alexis de Ganetzky	100	125
German Corps	Colonel Adolf Schiel	300	250
Italian Corps	Camillo Ricchiardi	300	100
Scandanavian Corps	Commandant Flygare	100	50
Total		1,850	1,025

the end of three years of war, there were still 20,000 men who came in from the veld.

The guerilla phase of the war saw a development of the Boer tactics, with the Boers reviving the cavalry charge by galloping at the Khakis in open order, firing from the saddle and enveloping the flanks. The shock reaction from magazine rifles had replaced the shock reaction from sword and lance. Groups of Boers also learned to assault positions in rushes covered by supporting fire from other groups.

In the early part of the war, the Boers were chivalrous opponents who treated prisoners well. They were so used to shaking hands among themselves that elated Boers sometimes shook hands with their prisoners (Hillegas, p. 83). They were devout Calvinists and avoided fighting on Sundays. 'Do your duty and trust in the Lord' was a typical injunction. They saw a direct parallel between the destiny of Old Testament Israel and the Afrikaner people. Both were chosen tribes overcoming their enemies in defence of a land given them by God.

In the later stages of the war, the behaviour on both sides became more brutalised. In particular the 'bitter-enders' detested the 'hands-uppers' who joined the National Scouts to fight on the British side. There was little mercy given to those whom the Boers regarded as collaborators.

THE BRITISH

At the end of the nineteenth century, the British Empire had an area of nearly 11 million square miles and a population of 378 million. It had an army with an establishment strength of 700,000 men – although the actual strength was much lower – who could be transported anywhere by a vast fleet. Great Britain could call on forces from many other parts of the Empire. But it was an army firmly entrenched in tradition and an army led by generals who did not at first realise how effective an opposition they had to face. Whereas the primary aim of the Boers was to minimise the loss of life, the primary British aim – at least initially – was to take the enemy positions by a frontal attack regardless of cost. The army of farmers had to be taught a quick lesson.

Almost every infantry regiment in the British Army fought in South Africa. An infantry battalion at this time consisted of eight companies, each company being divided into four sections. An appropriate wargames representation is one figure per section, making a battalion of 32 figures at full strength. Due to illness and casualties, it would never in fact be at its full complement. A wargames battalion of 24 figures is a reasonable size.

Units from the Naval Brigade fought alongside their army counterparts. One company of seamen (50 men) and three companies (190 men) of the Royal Marine Light Infantry fought in the frontline at Graspan, losing 15 killed and 79 wounded or 40% of their strength. An appropriate wargames representation would be two seamen figures and eight marines.

Actions such as Graspan illustrated the lack of training of the infantry as well as of the sailors and marines in taking cover. In one action, a squadron of South African Light Horse held a kopje all day, under heavy fire, without losing a single man. Nearby, two companies of British regulars held a hill under similar fire and lost 20 men, because they neglected to keep under cover (Fordham, p. 50).

The drill book did not cover the tactics for infantry to use in attacking an unseen enemy firing smokeless powder from invisible positions. Although most of the British units adapted fairly quickly to the use of extended order, there were still examples of troops in close formation providing a sitting target for Boer sharpshooters. At Colenso, one brigadier did not believe in extended order and tried to keep his brigade in close formation (Barnes, p. 143). At Magersfontein the Highlanders were caught at 400 yards range before they had deployed from quarter-column. Often the senior officers did not realise the range at which Boer rifle fire was effective until they learned by hard experience. As the Boers were hidden in trenches and held their fire until the Khakis had unwisely committed themselves, the distance to the enemy was often not evident in any case.

The British standard attack procedures called for an advance in column to within half-a-mile of the enemy before deploying. At that point one or two sections from each company were supposed to deploy into a firing line at extended intervals. In a series of rushes, covered by supports and reserves, the firing line would reach to within 300 yards of the enemy, when they would halt and fix bayonets. After three more rushes and a breathless charge, they would then overwhelm the demoralised enemy. This procedure had no hope of success against the Boers. Unless they were waiting in ambush, the Boers would already be scoring hits before the column had deployed and a frontal attack, no matter how gallant, would not get near its objective.

In the initial actions of the war, only one-tenth of the British forces engaged were mounted. This was a serious mistake in view of the mobile nature of the fighting. The Dominions had even been asked to give priority to the supply of infantry units. When the mistake was realised, the Empire was scoured for horses to equip mounted infantry units.

Of all the British cavalry regiments, only the 4th Dragoon Guards, the 2nd Lancers, and the 4th, 11th and 15th Hussars did not eventually fight in South Africa. A cavalry regiment of four squadrons, each consisting of 5–6 officers and 120–130 troopers, can be represented on the wargames table by 20 figures. The day of the cavalry charge was not yet completely over and actions such as Elandslaagte and the four-mile gallop, under fire, of French's Cavalry Division to bypass Boer forces during the relief of Kimberley showed the continued worth of regular cavalry. However, these successes were probably counter-productive in terms of preparing the British Army for the conflict that was to come in 1914.

A cavalry division usually consisted of two brigades, each composed of two or more regiments having two, three or four squadrons divided into two, three or four troops. A troop at full strength consisted of 20 sections of fours. A cavalry division of six regiments in two brigades brought up to strength for service abroad was made up of nearly 300 officers, 5,900 men, 12 guns, eight machine guns, 5,900 horses and over 300 wagons (*Soldiers of the Queen*, issue 33). There were more men than horses as some sections, for example the Field Hospital, were not all mounted.

The City of London Imperial Volunteers consisted of two companies of mounted infantry, an infantry battalion and a four-gun battery. The original total strength was 1,275 officers and men. This was supplemented by a later draft of 150 men.

The interesting concept of 'light horseman' can be applied to units such as the Australian Light Horse, which comprised excellent horsemen trained to fight on foot – as opposed to infantrymen temporarily provided with increased mobility (*Savage and Soldier*, volume XIV, issue 1, p. 2). The Australian Light Horse could fight on horseback as well as on foot. Most units carried the bayonet and sometimes made mounted charges, firing from the saddle as they went in and then using the bayonet.

In a wargame, Australian Light Horse count less in mêlée power than regular light cavalry but more than mounted infantry. Dismounted, they count as light infantry with sharpshooting ability. Their assigned tactics are to charge as near as possible to the enemy, dismount, fire and then go in with the bayonet. Interestingly enough, towards the end of the war, the Boers also made mounted charges on a few occasions.

The role of the artillery was vital in South Africa. Current doctrine laid down that artillery should first silence the enemy's guns and then neutralise the position the infantry were to attack. But it was frequently impossible to identify the Boer artillery positions or rifle lines, because of the smokeless powder they used. For this reason, or because of the impetuosity of local commanders, the artillery sometimes fell into the trap of advancing too close to the Boer riflemen. At Colenso, ten guns had to be abandoned. The 1896 manual clearly stated that, when engaged with infantry, artillery, if it had the choice, should take up a range exceeding 1,700 yards. Good infantry should, it was pointed out, make artillery pay heavily for coming within rifle range, or under 1,200 yards. This is one of the occasions when playing it by the book would have been wise. Yet there were cases of guns being advanced – in the open – to a distance of under 1,000 yards from the Boers.

The difficulty of identifying targets led to several unfortunate incidents. At Magersfontein, the British artillery bombarded their own infantry. The same happened at Talana and the artillery compounded their error by not firing on retreating Boers, whom they mistook for the 18th Hussars as the Boers were wearing capes in the rain.

A field battery of six 15-pdr guns crewed by 90 gunners, supported by 80 drivers and others, can be represented by one or two gun models, eight artillery figures, six horses and two wagons; while the slightly smaller complement of a horse battery of six 12-pdr guns can be represented again by one or two gun models and two wagons but with seven artillerymen and seven horses.

A field company of engineers can be represented by one or two mounted figures, six or seven dismounted figures and a wagon. The establishment of a mounted infantry company in 1898 was five officers, 137 men and 142 horses, which can be represented as five or six mounted figures. The maxim gun section of a mounted company consisted of one officer, 16 men, 18 horses, two maxim guns, two limbers and two wagons – too small to be represented. However, if we assume three maxim guns are working together, we get two crew figures, two horses, one maxim gun, one limber and one wagon in 1:25 wargames scale.

The formal organisation of the British forces changed completely in the guerilla phase of the war. Columns of mounted infantry were set up to sweep the veld and drive the Boer guerillas towards the blockhouse lines. It became a war of encounter, ambush, manoeuvre and skirmish.

There is not space here to list the large number of units that fought on the British side in the war. A comprehensive list of British units engaged in the war is given in Volume Two of *With the Flag to Pretoria*. Non-British units are listed in Gordon, p. 263. This list gives the strength of the 57 Australian units and 10 New Zealand contingents that fought in South Africa; it also lists the 110 South African Town Guards and all the Imperial Yeomanry companies.

2 SPECIAL FEATURES OF THE WAR

Capturing the flavour of the Second Anglo-Boer War requires a number of special features to be taken into account. Without these, the distinctive characteristics of the war cannot be recreated. The following section therefore gives a short summary of the special points that have to be addressed in writing effective wargame rules for the period.

Accuracy of Fire

It is generally held that the Boers were much better sharpshooters than the Khakis. However, even the Johannesburg commando, composed of civil servants, legal clerks and shop assistants (Reitz, p. 20) proved deadly. A thousand Boers, discharging their magazines from cover, could deliver a lethal storm of 5,000 bullets on a flat trajectory that swept the ground before it. The Boers could even fire with effect from horseback, although the effect should perhaps be reduced to, say, a quarter that of dismounted fire. One weakness of the Boers was to fire high and initial fire effect should be reduced by one half. On the other hand, when the range is known exactly from previous measurements by the Boers, the fire effect should be increased by a factor of 1.5. Often the Boers laid out white stones at predetermined ranges. Individual Boers could of course be superb shots: Piet Boueer of the Pretoria Commando hit a British soldier in the open at 1,400 yards (Hillegas, p. 310).

Armoured Trains

The railways were vital for bringing up troops and supplies but were vulnerable to attack. Armoured trains were therefore developed, usually armed with a small gun. The trains were of course still limited to static lines and were relatively easy to disable by destroying the lines.

The 1899 armoured train was mostly used for scouting and consisted typically of a truck armoured with iron boiler plates leading the engine and tender, with a second armoured truck at the rear. The trucks had slits for the occupants to fire out but as they were open at the top were very vulnerable to casualties from plunging fire. Sometimes a flat car carrying an artillery piece was located between the tender and the rear truck. If there was time, this would be unloaded before coming into action.

The 1901–02 armoured train was a larger and more technologically advanced affair. It was used for line repairs and for supply of beleaguered outposts on the blockhouse lines. There might be as many as seven trucks in addition to the engine, variously equipped to carry a Maxim with a 360 degree field of fire, an artillery piece with a more restricted operation, a searchlight, half a company of infantry, supplies or obstacle-clearing equipment.

A Railway Pioneer Regiment was formed on 18 December 1899 to provide specialist experience in undertaking repairs to the railways. The recruits were mostly Uitlander refugees. Initially it consisted of a single battalion in three wings, each of three companies. The total complement was 1,035, comprising 32 officers, 987 men, eight staff and eight medical personnel. The regiment had a military as well as a railway engineering role and took part in the battles of Roodewal on 7 June 1900 and Zand River on 14 June 1900 (*Soldiers of the Queen*, issue 52, March 1988). It also had a police role, in which capacity some men were mounted. Two other battalions were added later in the war.

Balloons

Although the military value of a telephone link to an observation balloon was not yet realised, balloons were used to observe enemy movements. They were particularly prominent in the defence and relief of Ladysmith and were also used at Colenso, Modder River and Paardeberg. Their observations helped to determine the position of otherwise undetectable guns firing smokeless powder. The balloons were a symbol of superiority of the British and upset the Boers, who directed much rifle and shell fire at them. However, the balloons were generally safe at high altitude. They were most vulnerable near the ground and the balloon teams soon became highly skilled at rapid hauling down and letting up. As far as is known, the balloons were not used for dropping explosives.

In 1900 the Balloon Corps in South Africa consisted of one section, with three officers and 31 NCOs and men. By 1901 the establishment had been increased to three sections, with eight officers, 173 NCOs and men, and 36 horses. Balloons were no longer required in South Africa after 1901 and the 1st and 2nd Sections became the 3rd and 2nd Field Troop RE respectively;

while the 3rd Section was transferred to the Railway Department.

Barbed Wire

This was not used directly to any great extent in the formal stage of the war but barbed wire fences were used to delineate the boundaries between farms and sometimes obstructed the movement of troops. At Modder River the Boers threw rolls of barbed wire into the water to discourage the Khakis from crossing above or below a road bridge. Barbed wire was used to link the blockhouses in the guerilla phase of the war.

Blockhouses

These were a key factor in the operations against the guerillas. The blockhouses varied in size but typically might have an internal diameter of 12 feet, a pair of walls sandwiched with shingle to check bullets and narrow firing slits. Around the outside ran a ditch in which the night sentry – often an African – could patrol in relative safety. Fordham (p. 185) indicates that 20 experienced men could build three blockhouses in two days, which seems a high work rate. Each blockhouse was supplied with 8,000 rounds of ammunition and enough food and water for 14 days. Sometimes searchlights were mounted to sweep the veld. An infantry battalion had to garrison up to 35 of these miniature forts. Thus a subaltern was responsible for three or four forts and a captain for ten or so. Each officer was linked by telephone.

Disease

Far more British troops died in South Africa from enteric fever (typhoid) and other diseases than from enemy action. Wargames units should be kept below full effect to allow for a proportion of troops being in hospital. In all, some 13,000 officers and men died of disease and the effect of sickness on some units must have been dire.

Dum-dum Bullets

Both sides on occasion used expanding bullets that 'mushroomed' inside the body and left a gaping wound on exit. A large stock of dum-dum cartridges was captured by the Boers at Dundee and these were used in subsequent battles. Ordinary Mauser cartridges were also converted by simply cutting off the point of the bullet. Whole boxes of Boer ammunition found at Inniskilling Hill near Ladysmith had the top cut off to expose the soft core and four slits scored down the side of each bullet. On the other hand, cartridges found at the same site covered in bright green slime were probably not poisoned, as claimed at the time, but protected by wax (Churchill, p. 498). Occasionally cartridges used for shooting big game were fired by mistake (Reitz, p. 136) and these had the same effect as expanding bullets. Many of these 'sporting' cartridges had apparently been supplied by Ely of London. In skirmish games, all bullets of this type score major wounds.

Dynamite and Mines
A few examples occurred where both sides improvised grenades out of sticks of dynamite and a fuse and these had some effect in attacking small fortified posts. During the siege of Mafeking, the Boers filled a trolley with dynamite, tied a fuse to it and pushed it down the slope into town. The dynamite exploded prematurely (Comaroff, p. 7). At Colenso, the Boers mined the road bridge but left it intact so that it could be blown up to cut off British troops once they had crossed.

Field Telegraph
The importance of telegraphic communication was not realised at the beginning of the War. Only 80 miles of cable were sent to the Cape with initial supplies. By the end of the War, over 18,000 miles of cable had been laid, which the Boers delighted in cutting. An experiment in wireless telegraphy, using equipment inter-cepted on its way to Kruger, was – sadly – abandoned as a failure. The Royal Engineers had special sections to lay the wires.

Heliograph and Other Signalling Techniques
The heliograph was a hinged mirror on a tripod which could be used to send coded messages on a cloudless day. The range varied according to the size of the mirror but even the smaller versions could transmit 30 miles and more. The 10-inch mirror was reported to have been used at ranges of over 100 miles. The Boers had one heliograph captured from the Jameson raiders and operated a small unit of heliographers under Captain Scheppers. Signalling by heliograph was used throughout the War, although it gradually came to be replaced by line communi-cation; it probably reached its peak efficiency during the War, with speeds of up to 16 words a minute being achieved.

Subsidiary methods of signalling used semaphore flags and oil-burning lamps. At Spionkop the only lamp went out because of lack of fuel and vital messages could not be sent in the dark. Another method of communication was by pigeon post, which was particularly useful in sieges such as Mafeking. Apart from anything else, the food value must have been appreciated. Such communication methods probably need be taken into account only in map movement to determine which force can communicate with another.

Laagers

The moveable Boer laager consisted of wagons drawn up in a hollow square or circle, the disselboom or pole of each wagon being pushed under the body of the one in front. The openings were filled with bush or a wooden framework and the wheels from wagon to wagon were lashed with the rein chains.

Lyddite

This was introduced in 1898 to replace gunpowder as the bursting charge of common shell for all breech-loading guns of over 4.7-inch calibre. It was so-called as it was invented at Lydd in Kent. It was packed in special yellow-painted shells which had fuse holes in the top and were cast thick at the base in order to lessen the effect of the gun's discharge on the explosive content of the shell. It gave off an acrid yellow smoke and was supposed to terrify the Boers but in fact against dispersed troops, Lyddite seems to have had no more effect than gunpowder, as the explosion was concentrated within a small area. The fragmentation of bodies was, however, appalling.

Mounted Infantry

Each infantry battalion in the British Army had included a mounted infantry detachment since about 1888 for rapid reinforcement of threatened points, occupying positions for rifle companies to take over or acting as scouts. On occasions, detachments were formed into companies. At the beginning of the War, the importance of mounted infantry was not realised and it was infantry units that were sought from the colonies. By the end of the War the use of mounted infantry, supported by mobile artillery, had been shown to be the decisive factor in most actions.

Photographic Section

One officer and one NCO, mounted on bicycles, were employed with 1st Cavalry Division to take panoramic photographs. This would be a delightful addition to a wargames unit. Early motion pictures were also taken during the war by W K L Dickson of the American Biograph Company and J B Stanford, an amateur newsreel cameraman. These were short and of poor quality but provided a taste of things to come: they were shown in the intervals of live performances at music-halls. Posed anti-Boer pictures were also shot on Hampstead Heath.

Ruses de Guerre

The much reputed abuse of the white flag by the Boers may have been as much due to the general confusion inherent in any battle as to a desire to draw their opponents into the open to be shot at. It also happened in reverse: Du Plessis was shot accepting the surrender at Reddersburg. The Boers were, however, very wily and were not averse to pretending to flee or to deceiving the British by wearing khaki uniforms. The civilian population in a 'neutral' village might lay out a meal with the best china and a white tablecloth for an occupying patrol so that Boer sharp-hooters could be given time to surround the patrol. On one occasion the Light Horse got their own back by dressing as Boer women; before they knew it, several unsuspecting Boers were trapped (Wilson, Vol. III, p. 348).

Because of the prevalence of Boer spies among the local population and even in the volunteer units, British generals on occasion spread misleading information via the press. The real plan was intended to be kept secret and various misleading orders for troop movements were given over the telegraph wires and then cancelled in code. This seems to have confused the British as much as the Boers, since communications were poor to start with and this complication must have caused a few apoplectic fits. On one occasion, Roberts gave false information about troop concentrations to a war correspondent, with strict instructions to keep it to himself. When the true line of attack became apparent, the correspondent, who had – needless to say – already sent off his despatches, complained to Roberts about his unfair and dishonest treatment.

Steam Traction Engines

A number of steam traction engines and trucks were sent out to South Africa under the charge of Lieutenant-Colonel J L B Templer. These were first landed in Natal and afterwards tran-shipped to Cape Town, where some were used to carry stores from the docks. The remainder were used at Kimberley, Bloem-fontein, Johannesburg and Pretoria. At these centres, where coal

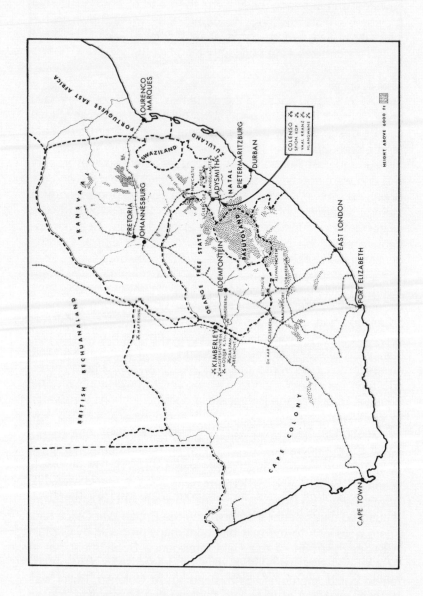

Warfare of the Future: The Traction Engine Mounted Infantry in Action (From *The Daily Graphic* 2 December 1899).

and water were readily available, the engines proved a valuable adjunct to animal draught; but, owing to the absence of fuel, they could not generally be used on the line of march or to haul supplies to bodies of troops more than 20 miles from a coal depot.

Terrain

The Boers had the great advantage of knowing the local terrain. Indeed, sometimes they were fighting on their own farms. Even in unfamiliar terrain, they were able to use excellent maps prepared by a corps of experts employed by the Transvaal government. The Boer maps of the environs of Ladysmith were a hundred times better than anything the British War Office had. There were inch-to-the-mile blueprint maps prepared by Captain A Kenney Herbert of the Field Intelligence Department but at Spionkop, these were more of a hindrance than a help. The terrain could, moreover, be misleading. At Colenso, Buller was deluded by a line of trees into thinking that Hlangwane was on the opposite side of the river. As a rule of thumb, the British side in a wargame should be regarded as being completely out-scouted by the Boers and to know the terrain far less well.

The South African veld consisted of a brownish-coloured landscape broken by high ant-hills and a few clumps of mimosa and prickly pear, with a backdrop of rugged kopjes and a strip of bush marking the course of a river bed. By the end of the war the British had embellished the landscape with 3,700 miles of barbed wire barricades and 8,000 blockhouses.

Trenches

These were of two types. The well prepared trenches used by the Boers at Modder River and Magersfontein provided effective cover. At Magersfontein over 1,000 yards of trenches were constructed. Being sited at the foot of the hills rather than on top, the trenches enabled the Boers to rest their weapons and to fire parallel to the ground and the effect on the advancing British troops was lethal. The most conspicuous Boer trenches were often dummies set up to deceive the Khakis. The other type was the hastily constructed trench used for example by the British at Spionkop, which did not provide good cover, particularly as in this case the trenches were not on the crest of the hill and were exposed to enfilade fire.

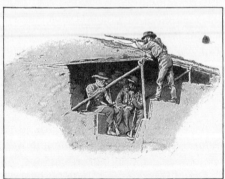

SHELTER TRENCHES
These diagrams show how some of the Boer trenches would look if cut through. Examples of both kinds were found at Magersfontein. The British in the besieged towns used similar protections. Often sandbags were added for further protection.

3   UNIFORMS OF THE BRITISH ARMY

The Second Boer War was the first major war in which the British Army wore khaki. It is possible that some of the units arriving in South Africa in their Home Service dress of red coats, blue trousers and white helmets may have continued to wear this in action but the predominant colour for painting wargames figures should be khaki drab. Many troops arrived from India in their khaki drill but this was found to be too thin for wear in South Africa and was replaced by a thicker drab serge, which became British Army standard issue.

A typical kit for an infantry private consisted of a white cork foreign service helmet with khaki cover, serge tunic with stand-and-fall collar and patch pockets, serge trousers, puttees, flannel shirt and body belt, socks, vest and drawers. The 1888 Slade Wallace equipment was made of buff leather whitened with pipe clay except in the case of rifle regiments, which had black leather. All metal parts were brass. The haversack was made of white canvas (black for rifle regiments) and the valise was of black lacquered canvas reinforced with black leather. The oval water bottle was covered with grey or blue felt. The mess tin was covered with black oilskin. In full marching order, a rolled greatcoat was also carried (*Soldiers of the Queen*, issue 45).

The pipe-clayed Slade Wallace equipment was soon found to be too conspicuous a target for the Boer sharpshooters and was commonly dyed, painted or stained a dirty brown colour. At Talana officers of the King's Royal Rifles wore a black crossbelt with the silver badge of their regiment over a khaki uniform and this provided an excellent aiming point for the Boers. These badges and head-dress badges, collar badges and brass shoulder initials were subsequently removed. As officers in many cases carried rifles rather than revolvers, little remained to distinguish them from their men except for their prominence in action. After these precautions had been taken, the number of officer casualties as a percentage of total casualties fell from 14% in the early battles of Talana Hill and Elandslaagte to an average of 6% – still twice as high as the ratio of officers to men.

The one distinguishing feature of officers in many cases was a puggaree round the helmet: headquarters staff officers had a crimson twist, divisional staff officers a dark blue twist, brigade staff officers a dark red twist, cavalry staff an all-red puggaree, artillery staff all blue, engineers' staff red and blue, cavalry officers a green twist, Medical Corps officers a puce-coloured patch on the puggaree and Army Service Corps officers a dark blue patch.

Helmet patches were also worn by privates: sometimes these were of the same colour as the twist of the officers but often they were more complex. The Royal Scots, for example, wore a section of red, white and green squares on the puggaree; while the Lancashire Fusiliers adopted a white embroidered grenade above the letters LF on a red patch. The patches were stitched to the helmet. Sometimes a shoulder strap bearing the regimental title was used as a patch, usually on a red square. The infantry patch was usually worn on the left side of the helmet but sometimes on both sides and occasionally in front.

The 17th Lancers wore their white-metal skull and crossbones badge but many other cavalry units had no distinguishing mark. Medical orderlies wore a white brassard with a red cross on the left arm and a small white disc with an embroidered red cross and a red circle on the upper right arm. Despatch riders wore a blue and white brassard. The Medical Corps wore a purple square on the helmet. It is a good idea to represent outsize coloured helmet patches on your wargames figures as it helps to pick out individual units from the khaki mass.

The main infantry units not to surrender entirely to khaki were the Highlanders. They wore a khaki cover to the helmet, a khaki tunic with the skirts cut away in front, and khaki gaiters or spats. Until the action at Modder River, they retained their tartan kilt, sporran and diced hose. After this action, a khaki cover was provided to cover the sporran and the front of the kilt. The problem of exposure to the sun still remained: at Magersfontein, when the Highlanders had to lie out in the open, the backs of their knees became so blistered they could hardly walk.

Even the naval units – the sailors and the marines – wore khaki but the former were at least distinguishable by their yellowish tan straw hats with a black band. The bluejackets are reputed to have hated their hot and uncomfortable khaki dress.

The Royal Artillery wore dark blue or khaki puttees, khaki jacket and trousers and a helmet with a khaki cover. Exceptionally, the NCOs wore gold embroidered rank badges on blue cloth. The wooden parts on the guns and gun-carriages were painted a yellowish-khaki. The General Service and ambulance wagons were also painted in this colour. The latter had a red cross in a white circle on both sides of the wooden body and of the canvas cover.

The cavalry uniform was also predominantly khaki and consisted of a slouch hat or sun helmet with a loose khaki cover, a khaki jacket with steel shoulder chains, khaki breeches, khaki puttees, a brown leather bandolier, rolled blue cloak and brown boots. Officers favoured a khaki helmet, leggings with spiral straps, a scabbard of brown leather and a Sam Browne belt, which had two shoulder braces when worn with a revolver; they sometimes wore the small folding field cap, usually in dark blue with gold furnishings. The horse kit was light brown or tan, except for the white picketing cord and mess tin of white tinned iron. A plain picketing peg was strapped behind the brown carbine bucket of dragoons. Other ranks in the hussars and lancers wore no waistbelt. The sword was carried in a leather loop on the shoe case attached to the saddle. The red over white lance pennon was not used in action.

The uniform of the City Imperial Volunteers, who set out from London 1,300 strong, was olive drab, with a grey-green slouch hat turned up on the left. Equipment was of brown leather except for a tan strap over the left shoulder and black boots.

The uniforms of the Colonial troops were slightly more colourful. The New South Wales Lancers wore brownish-khaki uniforms with greenish-black cockfeathers in their slouch hats and their insignia on the upturned brim. Shoulder straps, collar, pointed cuffs, plastron, hat band and trouser stripes were in red, set off by a yellow waist sash with two red stripes. The shoulder belt was white, the buttons and hat badge of white metal and the boots and gauntlets of brown leather. The saddle cover was of black sheepskin trimmed with red dragon's teeth. The mounted infantry were generally distinguished by green trimmings but the Victoria Mounted Rifles had purple/maroon facings with white piping, khaki jacket, light khaki pants, grey or light khaki hat turned up on the right side and brown belt. Some Australian units (the New South Wales Mounted Rifles, the 1st Australian Horse and the South Australia Lancers) wore the khaki sun helmet rather than the slouch hat.

Finally, the Cape Mounted Police of Cape Colony wore a dark blue tunic, white or light khaki breeches and a white helmet. The sword belt was white, the boots black and all leather equipment brown.

City Imperial Volunteers on scouting and patrolling work – crossing a small drift.

4   WHAT THE BOERS WORE

For the most part, the burghers fought in their everyday working clothes of dun-brown and black material. Typical dress was a dark coat, corduroy or moleskin trousers, a wide-brimmed hat and home-made shoes. It was not unknown for Boers to use umbrellas and shawls (Hillegas, p. 79); while De Wet had a briefcase to carry heliograms and other military papers. A style of dress amounting almost to an officer's uniform (Reitz, p. 25) was a black claw-hammer coat and a semi-tophat trimmed in crêpe. It is not true that no uniforms at all were worn outside the artillery: at the beginning of the war, the Johannesburg and Pretoria commandos were dressed in specially made khaki uniforms. In the guerilla phase of the war many Boers also wore khaki uniforms taken from British prisoners and some Boers captured in khaki were shot, particularly where they had used this as a ruse de guerre to deceive the British.

The Boer artillery was a disciplined and uniformed force. The Staatsartillerie of the Transvaal had a full dress of blue frock coat with brass buttons and green facings and a white helmet with white horsehair plume; while service dress consisted of a light mouse-coloured drill jacket with five brass buttons and royal blue standing collar and shoulder straps, Bedford-cord drab breeches or trousers, jack boots and a felt bush hat turned up on the right.

The Staatsartillerie of the Orange Free State wore a blue tunic in full dress, with black collar, cuffs and shoulder straps piped in orange, and black breeches or trousers with an orange stripe. Head-dress was a Prussian-type peaked cap or spiked helmet with an orange and white plume. The fatigue dress seems to have been a pale grey uniform of German pattern piped orange, worn with a round peakless cap. The field uniform, which is the one of most interest to wargamers, was a simple affair of whitish-brown colour, with no piping or braiding. The officer's tunic had a small standing collar; that for other ranks had a large turned down collar. The head-dress was a bush hat with the badge of the Orange Free State.

Another uniformed force, the Transvaal Police Force, also wore blue serge, with white helmets: the mounted elements wore

The author en route to a
wargame carrying a
briefcase of urgent
heliograms.

WHITE ENSIGN (H·M·NAVY.)

ADMIRALTY TRANSPORT

UNION JACK

ORANGE FREE STATE

TRANSVAAL

THE BRITISH   AND BOER FLAGS.

The white ensign (having a red St. George's cross and the Union in the corner) is used by Her Majesty's ships; transports fly the blue ensign with the golden anchor of the Admiralty. The Transvaal flag has three horizontal stripes, red, white, and blue, with a green vertical stripe next the staff. The Free State flag is (or rather was) the only flag of any state having orange as one of its colours; it was striped alternately orange and white, with red, white, and blue stripes occupying the first quarter.

The British and Boer flags.

Bedford-cord breeches with stripe and jack boots. There seems to have been a more informal dress of khaki bush hat, dark brown corduroy jacket, grey or khaki breeches, brown leather belt and brown boots.

A touch of colour can be added to wargames units of Boers by adding flags. The vierkleur of the Transvaal consisted of a green vertical stripe with three horizontal stripes in red, white and blue. The flag of the Orange Free State had a small square in the top corner of red, white and blue stripes, against a background of three orange and two white stripes.

It is interesting to note that when Britains Limited introduced Boer troops into their range of lead soldiers at the turn of the century, they did so with figures that had hats painted black – both for the infantry (set 26) and the 'cavalry' (set 6) – to show that they were the 'baddies'. The infantry were incorrectly equipped with bayonets.

5   THE GENERALS

In the formal battles of the war, the British generals were personally brave and gentlemanly but they tended to be bull-headed and deficient in the mental agility necessary to improvise quickly in an emergency. When they did devise good battle plans, they lacked the confidence to keep to them when difficulties arose (Barnes, p. 142). Reconnaissance was faulty and they had to work with poor maps from which they made false deductions. The performance of Sir Redvers (Reverse) Buller has come in for particular criticism from military historians, although at the time his men were said to worship him. At Colenso there is little doubt that he abdicated all responsibility and allowed the attack to peter out. He himself admitted he was a better second-in-command than a leader.

Most of the British generals have to be classed for wargaming purposes as below average: they still favoured frontal attacks and failed to put in decisive flanking movements or to get behind the Boers to cut off their retreat. Although some of the generals

The wagon used in South Africa by Roberts, photographed at the Museum of Army Transport in Beverley, North Humberside.

British Foreign Service GS wagon made by Dorset (Metal Model) Soldiers Ltd.

began to learn as the war progressed, only Roberts showed a consistent understanding of the proper battle tactics. He should be classed as above average. Kitchener's tactics were nothing less than the charge of a bull at a gate.

Not the least of Roberts' innovations was the weeding out of poor senior officers in his command. He sacked, for incompetence, no less then five generals of division, six brigadiers of cavalry, 11 commanding officers of cavalry regiments and six infantry colonels. This seems to have been a fair measure: Buller's comment on one of his subordinates (who was lame) was that he had two legs but no head; while the troops' nickname for Sir Leslie Rundle was 'Sir Leisurely Trundle'.

The British generals also showed a tendency to neglect overall command in favour of taking personal charge of an artillery battery or an infantry company at what they thought was a critical time. This is plus two on morale for the unit concerned but minus two for generalship.

In contrast, the Boer fighting generals were highly effective. Apart perhaps from Joubert, they should be classed as above average. De Wet, Botha and Smuts were outstanding. There is a marked parallel with the confederate army in the American Civil War, which also had inferior numbers but superior generals. However, under Joubert's lack-lustre leadership, the Boers lost the initiative in Natal by becoming bogged down in the siege of Ladysmith, when they could have headed for Durban and the coast with very little opposition. But in the guerilla phase of the war, the mobility and skill of the Boer generals was countered only by ruthless action on the part of Kitchener.

6 WEAPONS

THE LEE-ENFIELD RIFLE, USED BY THE BRITISH TROOPS.

The cartridges B are placed singly in the magazine A, from which a spring at the bottom forces them upwards till one of them enters the breech, when it is pushed forward by the bolt D into the chamber C and fired. The withdrawal of the bolt ejects the spent cartridge. There is a slide which, when required, cuts off the magazine and allows single cartridges to be used. F is a cleaner and oil-can carried within the butt.

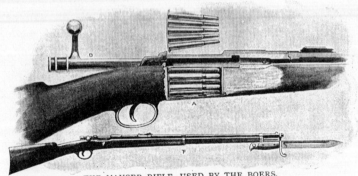

THE MAUSER RIFLE, USED BY THE BOERS.

The cartridges E are carried in a holder, from which, by one pressure of the thumb, they are released and dispose themselves in proper order in the magazine A. They are pressed upwards by a spring B, and forced, one at a time, into the chamber C by the bolt D. The rifle is sometimes provided with a bayonet, but this the Boers do not carry.

The Lee-Enfield rifle used by the British troops (*top)* and the Mauser rifle used by the Boers (*bottom*).

The main infantry weapons were the 1888 Lee-Metford 0.303-inch calibre rifle with bolt-action breech and the 1899 Rifle Magazine Lee-Enfield Mark I with the same calibre. The first Lee magazine held eight rounds of black powder ammunition but this capacity was increased to ten rounds when smokeless cordite ammunition was introduced. The magazine had to be loaded cartridge by cartridge and once the reserve of rapid fire had been used, there was no way of recharging the magazine quickly. The Lee-Metford was sighted to 2,500 yards and the Lee-Enfield to 2,800 yards; but effective range would have been about half these distances.

In contrast, the 0.275-inch calibre Mauser rifle of 1898 Gewehr pattern, of which 70,000 had been ordered by the Boers, could be recharged with a magazine clip of ammunition containing five cartridges. The Boer rate of fire was therefore significantly higher than that of the British infantry, regardless of the individual prowess of the firer. The Mauser also had the advantage on range: the sighted range was 3,274 yards. Again the effective range would have been about half the extreme range. The Boers frequently scored hits at 1,200 yards. At the beginning of the war, they had a plentiful supply of ammunition and carried 200 rounds apiece on their person: by the end of the war, the guerillas were reduced to trailing British troopers to pick up cartridges that had been dropped.

The Boers had another weapon apart from the Mauser – the spade. They expressed a preference for digging a trench rather than a grave. The British learnt this lesson at terrible cost.

The cavalry regiments which went out to South Africa at the start of the war were equipped with a Martini-Enfield carbine or the Magazine Lee-Enfield Mark I carbine and either the 1885 or 1890 pattern of sword. Later reinforcements were issued with the 1899 model, which was heartily disliked because of its 33-inch blade. This was suitable for neither cutting nor thrusting. The Martini-Enfield carbine proved very ineffective compared with the Lee-Enfield of the mounted infantry and was completely outclassed by the Mauser. The lancers were effective at Elandslaagte and, thereafter, the Boers reserved a special hatred for them. The lances were sportingly withdrawn and infantry rifles issued to all cavalrymen.

In 1899 the Royal Horse Artillery was armed with the 12-pdr breech-loading gun of 1894. The Field Artillery had the 15-pdr gun and the 5-inch howitzer. The Mountain Artillery gun, the famous 2.5-inch screw gun, was used early on in the war but proved a failure because of the black smoke it generated, which gave away its position.

The Royal Navy made an extremely important contribution to the effectiveness of the British artillery in the first six months of the war. Captain Percy Scott improvised field carriages for four long-range 12-pdr guns from *HMS Terrible*. These, and two ox-

A 5-INCH HOWITZER, AS USED BY THE BRITISH AT MAGERSFONTEIN.

Partly in section, showing the hydraulic buffers and the apparatus for raising and lowering the muzzle. This weapon can be fired at an elevation of forty-five degrees; it is intended to throw its shell high into the air so that it shall fall within the enemy's earthworks or other defences.

A 5-inch Howitzer as used by the British at Magersfontein.

Sir W. G. Armstrong, Whitworth, & Co.]

12-pounder gun and limber.

THE VICKERS-MAXIM 1-POUNDER.

Used by the Boers, and variously called by our men "Pom-pom" and "Ten-a-penny."

The Vickers-Maxim 1-pounder.

drawn 4.7-inch naval guns, helped to overcome the superior range of the Boer artillery and saved Ladysmith.

During the siege of Mafeking, three strange guns were pressed into service: Lord Cecil, a 4.1-inch gun with an extreme range of 8,000 yards that fired 255 28 lb shells during the siege; Lord Nelson, an 8 cwt naval gun of 1770 vintage; and The Wolf, a 4.5-inch howitzer made from a drain pipe and which Baden-Powell took home with him after the siege. Lord Nelson had a range of two miles with 3lbs of powder and 2,000 yards with 2lbs while The Wolf had a range of 2,500–4,000 yards.

FUZES USED ON BRITISH SHELLS

Fig. 1 is a percussion fuze, for exploding a shell on coming in contact with any solid object. The steel needle A forces the copper washer B against the detonating composition C, exploding the pistol powder D, thus firing the charge in the shell through the aperture E. Figs. 2 and 3 represent the exterior and interior of a time-fuze; it is prepared for use by loosening the cap A, and turning the dome B until the index C is set at the required number of seconds, when the cap is again tightened. The safety pins D D are then withdrawn by means of the cords F F attached to their heads. On being fired, the centrifugal motion causes the detonating pellet, released by the removal of the safety pins, to press against a steel needle, which fires it and thus sets light to the quickmatch. The fuze composition runs round the channel marked G, behind the index, which regulates the amount to be burnt before exploding the charge.

Fuzes used on British shells.

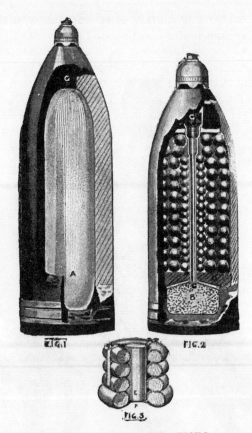

FIG.1   FIG.2

FIG.3

COMMON AND SHRAPNEL SHELLS.

Fig. 1 shows a "common shell;" on exploding, its fragments are hurled in all directions, and if charged with a high explosive, such as lyddite, the mere concussion caused by the release of the gases will cause death at a considerable distance. Shrapnel shell (Fig. 2) is filled with metal balls, which, on the explosion of the charge of powder, B, are flung forward in the face of the enemy. Fig. 3 represents the cartridge, of cordite, used in howitzers. Both shells are here shown with fuses attached, which may be arranged to fire the charge either after the lapse of a given number of seconds, or on impact. The fuse communicates with the charge through the channel, C.

Common and shrapnel shells.

The key to the war in South Africa: the Mounted Infantryman.

The Transvaal Staatsartillerie had, at the beginning of the war,
a battery of four 120mm Krupp howitzers, six 75mm Creusot
QF field guns, eight 75mm Krupp QF high-velocity field guns,

five 75mm Vickers-Maxim QF mountain guns and two Maxim-Nordenfeldt guns captured from the Jameson raiders. But their most useful armament was to prove four 150mm ox-drawn Creusot guns (the 'Long Toms') and 22 one-pdr automatic Vickers-Maxim pom-poms of 37mm calibre. The Boers dragged the Long Toms – the so-called 'fortress guns' – up places the British would have hesitated to venture with their mountain guns; while the distinctive sound of the pom-poms soon became familiar to British infantry pinned down on the veld. The Transvaal Boers had 31 Maxims of rifle calibre (0.303 and 0.45-inch). After the early British disasters, at least a dozen British guns fell into Boer hands.

The Staatsartillerie of the Orange Free State relied primarily on 14 75mm breech-loading Krupp field guns, supported by five 9-pdr Armstrongs and a 6-pdr Whitworth gun. It also possessed a 3-pdr Whitworth muzzle-loading mountain howitzer, a 37mm Vickers-Maxim QF gun and three Maxims of rifle calibre. The 75mm guns were older than the Transvaal pieces and used black powder.

The following table summarises the characteristics of the main types of artillery used in the war. The ranges quoted are for the extreme limit of fire and should be reduced to about half the distances quoted for effective fire with an observer, unless there are special circumstances such as firing at a fixed laager or strongpoint at a known distance. In any case, on the wargames table the firing distance is so long compared with normal movement distances that off-table siting of artillery pieces is required. The rates of firing quoted are realistic rates designed to take account of the need to conserve ammunition. Two batteries (12 guns) in the open and under Boer fire at Colenso fired 1,000 shells in one hour, ie 1–2 rounds per gun per minute. In the same action, two 4.7-inch naval guns fired 160 rounds but were in action for longer and were not so exposed to rifle fire. The rate of firing was of course significantly higher for short periods: the figures quoted are simply averages over the period of a battle.

There were rapid improvements in British field gun range during the war, increasing from 2,000 yards *effective* range in 1899 to 3,200 yards in 1902. Case shot was superseded as shrapnel with the fuse set at zero did everything that case shot could. The aim was to develop 'one shell and one fuse'.

Some 6-inch and 9.45-inch Howitzers were also sent out to South Africa but they were not found very effective and were little used.

Gun	Calibre (mm/ins)	Extreme Range (yds)	Type of Fuse & Notes	Fire (rpm)	Shell (lbs)
Boer Creusot 'Long Tom'	155	11,000	percussion	1	94
Boer Krupp/Creusot field gun	75	6,000 8,500	timed percussion	2	15
Boer Maxim-Nordenfeldt	75	5,000		5	
Boer Vickers-Maxim the 'pom-pom'	37	2,750 4,400	timed percussion	10 10	1
British 12-pdr RHA		3,800 5,600	timed later timed	2 2	12 12
British 15-pdr RFA		4,100 5,900	timed later timed	2 2	14 14
British BL Howitzer (45 degree elevation)	5	4,900–8,000	range depends on elevation	1	50
British RML mountain gun (largely obsolete)	2.5	3,300 4,000	timed percussion	2 2	7 7
British naval field gun	4.7	10,000		1	45
British naval 12-pdr		7,000	'long' 12-pdr	2	12

7 THE THIRD FORCE

There was a third party to the war in South Africa, even though the British and Boers had tacitly assumed at the beginning of the struggle that it would be a 'white man's war'. In practice, British forces employed large numbers of black and coloured people of mixed race as transport drivers, labourers and guides. Africans also fought alongside the British at Mafeking, although Baden-Powell sought to deny this. Over 500 blacks took part in the defence of the town, including the 'Black Watch' under Captain MacKenzie. Others were employed in local defence, for example the Zululand Police, and later on in the war as nightwatchmen on the blockhouse lines. It is estimated that at least 10,000 Africans were fighting with the British Army as armed combatants by the end of the war. Over 100,000 blacks became involved in the struggle as scouts, spies, guards, servants and messengers, mostly on the British side. Churchill (p. 503) believed that this assistance was the reward for the pursuance of a kindly and humane policy towards the Bantu races and that the Boers had paid a heavy penalty for their cruelty and harshness.

In the guerilla phase of the war, groups of black scouts, up to 50 in number, came to be attached to all the British columns, under white intelligence officers. One of the most successful of these groups was that led by Colonel Aubrey Woolls-Sampson. During two weeks in December 1901, 300 of the 756 Boers captured by British forces throughout South Africa fell to the columns to which Woolls-Sampson's men were attached. However, it is wrong in a Boer War game to use regular 'askari' units, as did one game reported in the United States magazine, *The Courier*, Volume III, no 6.

The Boers also used black guides and servants called ag-terryers (after-riders). Some blacks were conscripted into Boer commandos, particularly in the commando of General Christiaan Beyers. But most of these may have been used as agterryers rather than combatants. The Boers deplored the British practice of arming Africans, even if only for self-defence, and they showed little mercy to any armed black who fell into their hands. Often the same fate met any unarmed black who was suspected of helping the British (Fordham, p. 144).

While the black chiefs in British Protected Territories were able to consolidate their paramountcy during the war, the average black inhabitants of the British Colonies and the former Boer republics found themselves considerably worse off.

8 CASUALTIES AND THEIR EFFECT ON MORALE

The morale of fighting units was a key factor in the Second Anglo-Boer War, as indeed it is in any war. An unusual factor was the option of surrender in unfavourable circumstances. This was not an option that the British Army had been accustomed to in savage warfare. The pressure on British troops of the rain of Mauser and artillery fire was enormous. There were a few instances recorded in private letters of general officers hitting, kicking and praying the men to go on, and, in some instances, threatening them with death, but to no avail. At Spionkop Brigadier Woodgate was reputedly shot at by his own men while he was attempting to keep his brigade steady. In view of the absolute indifference of some of the British generals to the losses sustained in frontal attacks in the open, it is surprising that there were not more incidents of this kind.

The censorship of the time would of course have kept news of failure of morale very quiet. Not only was there the issue of patriotic duty; there was also the issue of preserving regimental pride. Each of the Highland Regiments at Magersfontein reported that they had done well themselves, although they could not necessarily say the same for other regiments. In fact, it was poor generalship that was largely responsible for the heavy casualties and consequent rout. Methuen himself said that he doubted if the Highlanders would ever serve under him again.

Generally the morale of the British troops was good. They were told what to do by their officers and they did it. Unlike the Boers, they were not told about the tactical position. Churchill records with disapproval an instance of a soldier not knowing the name of his brigade commander, but it is doubtful if privates were expected to know anything that did not relate to their immediate environment. Individual initiative was not a commodity to be encouraged. Unthinking bravery was.

'The last cartridge' became a convenient fiction to explain the surrender of British troops, for example at Reddersburg. British troops did not give up because they were outshot or because they suffered heavy casualties: they lost heart only when they ran out of ammunition and could no longer continue the fight. Under

A French view of the British surrender at Reddersburg, from a contemporary trade card.

Kitchener's command, officers that were considered guilty of giving up their posts too easily were dismissed; and heavy sentences were passed on NCOs and men found guilty of cowardice or unjustifiable surrender. At Rhenoster in September 1900, an NCO was sentenced to death (commuted to ten years' imprisonment) and 16 other men were given three to ten years. In the same town in May 1901, another 13 men were given five to ten years, the NCOs receiving the highest sentences (James, p. 203). These incidents were, however, an exception to a general rule of steadfastness under fire.

Contemporary accounts of Colenso stress that as long as there was ammunition in the limbers, one of the 12 guns continued to fire at intervals, until at last there were only two gunners left. In the donga behind the guns crouched or lay a few unwounded and a large number of wounded men (*With the Flag to Pretoria*, *Vol. I*, p. 93). Two officers – an Australian Colonel and a Captain Herbert – were reported to have claimed that every officer or man had been killed or wounded. In fact there is a limit to what flesh and blood can take. More sober accounts (including Pemberton, p. 139) indicate that out of 84 artillery officers and men, only 8 were killed and 12 wounded. The Boers took 64 unwounded gunners prisoner and captured the nearly full second-line ammunition wagons. The counter-claim is that these figures are merely Boer propaganda. What really happened in incidents such as this is still a matter for emotive debate between the traditionalists and the realists (see *Battle* magazine letters column for December 1976 and February 1977). As far as the wargamer is concerned, the implication is that once a unit has suffered more than 20% casualties from Boer rifle and artillery fire, it should be more or less out of action as far as good morale throws are concerned.

In only a few disastrous engagements such as Spionkop did overall British casualties exceed 7% – and in this case the deciding factor in the evacuation of this site was probably the incessant Boer artillery fire. The following table indicates British casualties in the major actions of the war. The figures are approximate, as sources vary in detail, but the overall pattern is clear. Excluding prisoners – which are an effect of loss of morale rather than a cause – British casualties were usually between 2 and 7% of the forces in the vicinity of the battle. Individual units in the front line of course suffered severe casualties: 25% of the Gordons at Elandslaagte; 27% of the Lancashire Fusiliers at Spionkop, causing many of the survivors to surrender; 35% of the Black Watch at Magersfontein; and 60% of the

Battle	Engaged	Killed	Wounded	Captured	Total	% k&w
Nicholson's Nek (defeat)	1,200	52	100	954	152	13
Sannah's Post (d)	1,800	30	113	428	143	8
Reddersburg (d)	590	10	34	546	44	7
Elandslaagte (victory)	3,500	50	200		250	7
Magersfontein (d)	15,000	220	690	70	910	6
Paardeberg (v)	20,000	300	950	65	1,250	6
Blood River Poort (d)	700	20	24	240	44	6
Spionkop (d)	27,000	300	1,000	280	1,300	6
Talana (v)	4,300	50	205	250	255	6
Stormberg (d)	3,000	35	100	600	135	5
Ladysmith Defence (v)	13,500	220	400		620	5
Colenso (d)	19,400	140	760	240	900	5
Modder River (d)	10,400	70	395	10	465	4
Belmont (v)	10,000	55	245		300	3
Graspan (v)	8,000	20	140	10	160	2
Vaalkrantz (d)	15,000	35	335	5	370	2
Rietfontein (draw)	5,300	12	100	2	112	2
Willow Grange (d)	3,500	10	70	10	80	2

Seaforths in the same battle. It was this local butcher's bill that created the make-or-break situation rather than the losses to the overall force. But the overall figures do show that the typical wargame, in which half the figures employed are casualties by the end of the game, bears little relation to the historical reality.

Furthermore, the proportion of wounded to killed was about 5:1 and there was a good chance of recovery. Small-bore bullets did not cause severe wounds unless they hit a vital spot and, combined with improved medical attention, only 10–20% of the wounded died.

The flexibility of the Boer commando system, under which the burghers could fight as and when they wished, meant that the Boers could not be relied upon to stay during the critical stage of a battle. Nor were all Boers indefatigable guerilla fighters. Reitz (p. 50) complained that the commando to which he belonged received reinforcements of inferior quality, mostly bywoners or poor whites from the slums of Pretoria, who had drifted in from the country after the rinderpest epidemic of 1896. So debased by town life were the recruits and so little stomach for fighting had they, that their presence weakened the commando rather than strengthened it. However, towards the end of the war, when the hands-uppers had defected, there is no doubt that the Boers left would have been battle-hardened veterans of high morale.

Boer casualties tended to be low. Usually they were in the region of 1–6% of the total force engaged. At Colenso they were barely 0.5%. An action such as Elandslaagte, when 22% casualties were suffered, was a complete disaster, since the Boers sought at all times to reduce their own casualties to a minimum. They believed that 'God Almighty gave them their lives for some better purpose than to be shot like bucks in the open' (Intelligence Branch, p. 50). There were exceptions: the Carolina commando fought on at Spionkop until it had 60% casualties. The following table gives an approximate indication of casualties in various actions. The figure in the column headed *Total* excludes prisoners.

An analysis of 249 wounds suffered by British soldiers in early engagements of the war indicated injuries as follows: in the head, 19; face, 7; neck, 3; back and spine, 20; upper extremity, 76; lower extremity, 118; and other wounds, 6. All the wounds except eight were caused by Mauser bullets which left a clean hole that healed readily (Fordham, p. 34). The pattern would certainly have been different at Spionkop, as there would have been a much higher proportion of head wounds caused by Boer sharpshooters.

Battle	Engaged	Killed	Wounded	Captured	Total	% k&w
Elandslaagte (defeat)	1,000	65	150	150	165	22
Rooiwal (d)	1,700	50	150	50	200	12
Spionkop (victory)	4,000	85	250		335	8
Paardeberg (d)	7,500	150	300	4,000	450	6
Magersfontein (v)	6,000	100	200		300	5
Talana (v)	3,500	40	100	10	140	4
Belmont (d)	2,500	50	50	40	100	4
Modder River (v)	2,400	50	30	25	80	3
Vaalkrantz (v)	4,000	35	45		80	2
Reddersburg (v)	800	1	6		7	1
Blood River Poort (v)	1,000	1	3		4	1
Colenso (v)	5,000	7	22		27	1
Rietfontein (draw)	6,000	13	31		44	1
Holkrantz (d) Zulus	70	56				80

9 WARGAMES FIGURES AND RULES

There are many good British and Boer wargames figures available on the market these days. As addresses and ranges supplied change from time to time, specific details are not given here. The latest information can be found in reviews and advertisements in the hobby press or by visiting wargames conventions.

At 15mm scale, excellent ranges are produced by Miniature Figurines Limited (Minifigs) of Southampton and Peter Laing of Hereford. Laing's World War I Australian cavalryman makes a good colonial horseman, and he also produces a CIV cavalryman as well as more usual types. Minifigs has a wide range of Boer War figures, including ox-drawn wagons.

Jacklex figures sold by the Model Shop in Harrow give sufficient variety in 20mm scale to build up balanced forces of all arms: there are six Boers on foot, including a horse-minder; two mounted Boers; British infantry, Highlanders, troopers and lancers; and 20 equipment items, including a horse-drawn supply wagon, a 4.7-inch naval gun and an eight ox team with limber and crew, and a Boer voortrekker wagon complete with three walking settlers (a man, a woman and a girl). Also in this scale, Airfix Confederates can be combined with ESCI types such as the Zulu War British, if you prefer plastics to metal figures and do not mind a little work in converting figures.

Good ranges in 25mm scale are those of Falcon Miniatures, Lyzard's Grin and Ral Partha, all United States suppliers. The Falcon series includes a Boer command group, three Boers on foot, two mounted Boers and an artilleryman. Ral Partha has sets of Boers advancing on foot, Boers firing on foot, mounted Boers and a Boer artillery crew and mounted officer; while the British figures include useful mounted irregulars. The animation of these figures is excellent, the Boers looking delightfully scruffy, Lyzard's Grin specialise in equipment items, including a Maxim-Nordenfeldt 75mm gun, a 12-pdr gun and ox and mule wagons. In the UK, Tradition of Shepherd Street, London, do a limited range of British (line infantry and lancers) and Boers (three Boers on foot and one mounted).

In 30mm scale Tradition also supply figures from the moulds made by Edward Surén. The Boers and British are beautiful

Boer commandant with two kripvreters (Surén 30mm figures).

Stadden and Surén officers confer.

figures, full of individuality, and they fit particularly well into skirmish games, on a one-to-one basis. The Boer commandant in top-hat and long coat is superb.

For an H G Wells type of wargame, there are many 54mm figures made in the style of Britains and painted in gloss finish. One of the widest ranges is that of Dorset (Metal Model) Soldiers Limited of Mere in Wiltshire. Their Boer War range covers General Buller and two staff officers, Imperial yeomanry, field artillery, Cornish Light Infantry, a pontoon wagon with boat and boards drawn by four horses, City Imperial Volunteers, Gordon Highlanders, British mountain artillery, Boers on foot, a Boer machine gun with four Staatsartillerie crew, and General Louis Botha accompanied by a colour bearer and a scout.

As far as rules are concerned, there used to be a distinct lack of good rules for the Boer War period but recently several sets have appeared that offer interesting possibilities. There is even a set specifically for the guerilla phase of the war: called 'Kommando!', the rules were published in an article by Ian Drury in *Miniature Wargames* magazine for February 1988 (no 57). They are designed to cover small columns of troops clashing on the veld and are intended for use with 15mm figures. A casualty card system plays an important part in simulating the difficulty of seeing what effect your firing is having on the enemy. Each player draws a number of cards, depending on how much fire is being directed at his units. Losses are not revealed until your units move off after the firefight. Thus your opponent will only know if your men are in trouble if the result of a morale test becomes apparent. Your opponent can check your casualty cards after the event to see that you have played like a gentleman.

The two main sets of rules used for the colonial competitions held during national championships have been *Colonials* by J E Davies of Chariot Miniatures and the Newbury *Colonial Rules* by T J Halsall. Boer armies have done well under both sets of rules, only being outclassed by opponents using gamesmanship to field hordes of Mahdist light cavalry advancing in open order and using morale effects to drive off the Boers. This happened particularly in the 1985 'Roll Call' at Luton, when the Mahdist commander exploited amendments to the *Colonials* army lists to use a force of 90 A class cavalry, supported by four breech-loading guns. In the final of the competition, this combination was unstoppable by the Boers, as defending skirmishers lose minus six off all morale throws. Full credit to the Mahdist player for knowing the rules backwards but hardly an authentic army!

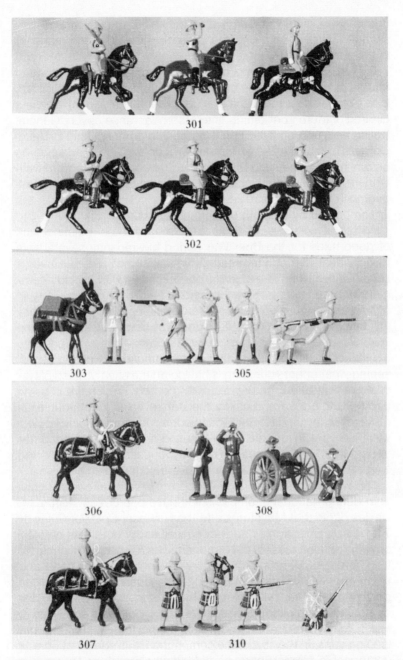

Boer War Dorset Soldiers. Examples include 301 Boer War Staff, 302 Imperial Yeomanry, 303 Ammunition Mules, 305 Cornish Light Infantry, 306 and 307 Engineers, 308 City Imperial Volunteers, 310 Gordon Highlanders.

The strength of the *Colonials* rules lies in the objectives they set, which prevent regular forces from using their superior firepower to remain static. For example, high points are awarded for reaching a target on the other side of the board. Another good feature is the use of rules for scouting. If one side is outscouted, it has to deploy all its troops except those in ambush; whereas the side with the better scouting ability has to set out only those troops that can be seen in the open. The army list for the Boers comprises A or B class veteran light mounted figures, all counting as marksmen, and A class veteran or average artillery crews with light field guns or pom-poms. The British are allowed A class average infantry, cavalry or artillery.

The Newbury rules are more detailed than *Colonials* and correspondingly more difficult to master. The effort is, however, well worthwhile. In competitions using these rules, players have been allowed to use only compatible armies and to choose a pair of forces totalling 600 points from a list of 1,000 points. In the 1972 event at Leicester, for example, the British (South African) list consisted of one general (100 points), 48 infantry (384 points), 12 lancers (144 points), 12 other cavalry (132 points), one machine gun (60 points) and crew (32 points) and two light guns and crew (200 points). The Boer list comprised one general (100 points), 75 mounted infantry (750 points), two light guns (at 50 points each) and crews (at 38 points each). There was a lack of provision in both army lists for medium or heavy artillery.

The Sebastopol, Sadowa and Sedan Nineteenth Century Warfare Rules by T J Halsall, J G Kew and A M Roth cover only the period 1830–1890 in their army lists but are nevertheless very useful as they include an appendix on special equipment such as observation balloons, signals equipment and smokeless powder. Boers are classed as C class irregulars at 5 points each, but can be upgraded to B class by paying a further point and to marksmen by paying 10 points. Horses cost three points. A British force of 1870–90 is allowed C class infantry, again which can be upgraded to B class by adding one point to the basic cost (ten points). Cavalry are A, B or C class; artillery are allowed only B class. Light or medium guns can be used. The Boers can include native servants as combatants on their side: this is historically accurate – indeed recent research has shown that 300 armed blacks aided the Boers at the Battle of Blood River in 1838, a point not publicised in Afrikaner literature. These rules were used for the 1989 15mm colonial competition at Dunstable and will also be used for the 1990 event, which will be based on the Indian Mutiny.

Imperialism by the Western Washington Wargamers in the United States is an authoritative set of rules specifically covering features of the Boer War such as pom-poms, trenches, forts, laagers and barbed wire. The Boers are given line morale until the British close, when they fall to militia morale. The chief author, Lynn Bodin, also edits *Savage and Soldier* magazine, an indispensable source of reference for the colonial wargamer, and has issued a series of notes on wargaming the Boer War.

Imperial Wars: Colonial Warfare Rules 1860–1900 by Mod Games of Sheffield includes a Boer army list of 40 plus B class irregular cavalry with breech-loading rifle at 12 points each, plus one point to upgrade to A class, one point for marksman status and one point to replace the BL rifle with a magazine rifle. Two BL rifled guns of 18-pdr class or less, with a B class crew, and two machine guns, with a similar class crew, are allowed. British infantry, cavalry and artillery are B class but can be upgraded to A class for one extra point per figure. The cost of a magazine rifle, above that for a BL rifle, is three points.

The Sun Never Sets is a short set of rules published by Dodo Publications of Sheffield. Boers are counted as regulars with a morale factor of nine, only one below British regulars. Wisely no attempt is made to cover all eventualities and it is left to the players to resolve unusual happenings. Some of the illustrations will be familiar to readers of the *Handbook for Colonial Wargamers*, a special publication of the Victorian Military Society.

Soldiers of the Queen: Wargames Rules 1879–1900 by D Elks and J G Stanyon is published by Tabletop Games. The army lists for the Boer War cover ten options, including Paardeberg, Colenso, Relief of Kimberley, Elandsaagte and Talana Hill. South African horse and mounted infantry move as irregulars but count as regulars for morale purposes. Provision is made for pom-poms, trenches, blockhouses, balloons and even traction engines. Trains are also covered: they move at a maximum speed of 24 inches per move so, unless they do any emergency stop, they are not on the table for long! A feature of the rules is the use of special event cards. For example, the one entitled 'Your Home is on Fire' causes all Boer units to test morale at minus two. Routs count as 'retire' but all units who get this must leave the board. Another entitled 'Ogilvy Rides Again' indicates that a dead British officer may be resurrected as the bullet bounced off his pocket watch. As this was once his grandfather's, he becomes annoyed and counts as a plus two officer.

As regards skirmish wargaming, *Skirmish Wargames 1850–1900* by M R Blake, S Curtis, I M Colwill and E J Herbert is now unfortunately out of print. The best alternative is *The Sword and the Flame* rules by Larry V Brom, which are widely used in the United States. The rules are professionally produced and contain some useful background information. A pack of playing cards is used to determine the order of moving and firing: if a red card is turned up, a British unit may move; if it is a black the initiative lies with the Boer player, and so on. Casualties using these rules tend to be rather high but they are fun to play.

As yet, role-playing games for the Second Anglo-Boer War are not covered by a published set of rules. However, Howard Whitehouse has produced *Science versus Pluck* or *Too much for the Mahdi* to provide tactical rules at a generalship level for actions at a brigade or divisional level in the Sudan campaigns of 1881 to 1899, and it would require relatively little adaptation to extend these to the two Boer wars. The rules consist of two booklets, an Officers' Pocket Book that gives players general background information and an Umpire's Handbook that is for the Umpire's use only and gives guidance on more detailed procedures. The success of a game using these booklets depends critically on the Umpire creating the flavour of the period and on the players acting out their part as particular British senior officers. This is an idea that deserves to catch on fast.

An invaluable source of information for the colonial wargamer is the magazine *Soldiers of the Queen*, journal of the Victorian Military Society. The Society has both a Wargames Study Group and an Anglo-Boer Study Group. Advertisements for the Society's annual fair appear in the military modelling press.

10   THREE SCENARIOS

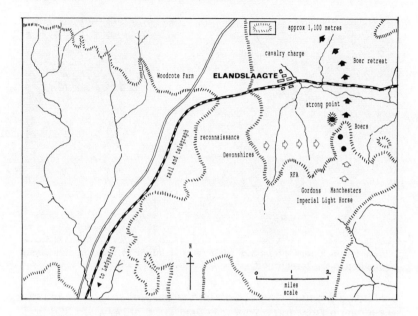

ELANDSLAAGTE: A SET-PIECE BATTLE

This battle has been chosen because it illustrates the neat planning of a set-piece battle, almost in the style of the Crimean War, but updated to cover use of the new technology – khaki, extended order, telegraph, armoured train and so forth.

The terrain over which the battle was fought has as its chief feature a horseshoe-shaped ridge south of the railway station at Elandslaagte, which lies some ten miles north-east of Ladysmith.

On 20 October 1899 a reconnaissance by the Imperial Light Horse and a battery of Natal Volunteer Artillery, with an infantry escort of four companies of 1st Manchesters, spotted the Boers at Elandslaagte but after turning a group out of the railway station, had to withdraw out of artillery range. Major-General John French used a telegraph tap to call Ladysmith and order up reinforcements by road and by armoured train. They did not arrive very promptly and it was not until late in the afternoon that all the reinforcements (five companies of the 2nd Gordons, the 1st Devons, the other half of the 1st Manchesters, the 5th

Dragoon Guards and the 5th Lancers, and the 21st and 42nd batteries of the Royal Field Artillery) were on their start lines. In all, French had about 3,500 men, 18 guns and six machine guns.

30mm Surén Boers preparing to defend a kopje; a rifle pit has been excavated in the foreground.

Holding a defensive position on the eastern arm of the horse-shoe was a Boer force variously estimated at between 350 and 2,000 men, mostly from the Johannesburg commando under Kock but including 100 Vrede men under De Jager and a contingent of 140 Germans and 70 Hollanders. The total strength was probably about 1,000 men. The Boers occupied two kopjes separated by a wide kloof. They had only two guns, in fact pieces captured from the Jameson raiders, and a Maxim. Their laager was sited on a flat stretch of sun-baked mud to the rear of a gap in the ridge. French possessed valuable information about the strength and disposition of the Boer forces, given him by the manager of Elandslaagte Coal Mine.

The battle began with artillery fire from the two RFA batteries at 4,400 yards. A squadron of the 5th Lancers and four squad-rons of the Imperial Light Horse drove off Boer skirmishers under Field-Cornet Pienaar from the ground to the south of the main Boer position and protected the right flank from marauding Boers; and the remaining cavalry guarded the left flank. While the artillery was in action, some Boers rode ostentatiously down the left slope of their position to try to draw the British horse under fire; but this ruse did not succeed.

Hamilton had meanwhile detrained his men behind the western

arm of the horseshoe and given them a pep talk. After a 30-minute artillery bombardment, the infantry advanced in extended order. There were three yards between each man in line and several hundred yards between each successive line. Seven companies of the Devons formed the frontal assault force, which would have to cross the two-mile gap between the arms of the horseshoe, while the Manchesters, five companies of the Gordons and the dismounted Imperial Light Horse went in from the south to turn the enemy's flank. The Devons had three companies in the firing line with 450 yards between successive lines of supports and reserves, so that the battalion had a depth of nearly a mile. At about 900 yards the British opened fire but the range was too great to have any serious effect. Thereafter they advanced by short rushes in extended lines. The Devons had to go to ground as sustained fire from the Boer positions began to tell, the only cover being ant-hills.

30mm Surén Boers under General Kock (in top hat) counter-attacking at Elandslaagte; the thunderstorm is about to break.

With the Boers hopefully pinned from the front, the flank attack went in. The Gordons suffered particularly as they dashed across some open ground. At the top of the ridge, wire fences further obstructed the attackers.

It was not until a thunderstorm burst over the battlefield that the flanking force gained the advantage. Advancing in extended order, the dismounted ILH supported the Gordons and the Manchesters. Resuming the frontal attack, up the escalade of steep, boulder-strewn slopes went the Devons from the east, bugles sounding. A final bayonet charge from the flank carried the day. The Boer laager was taken after an incident in which a white flag was hoisted and the Boers then fired on unsuspecting men who had left cover. The attack faltered but Hamilton was on hand to restore the momentum. The contingent of foreign volunteers lacked the Boer instinct for timely withdrawal. They were over-run. A few Boers fired on the British until they were only 20 yards away and then asked for quarter. However, most of the burghers ran down the reverse slope and mounted their waiting horses. It looked as though they were safely away.

But then disaster struck the burghers. From the west charged a squadron of the Dragoon Guards and a squadron of the Lancers that had been waiting at Elandslaagte Station. Several times they rode through the fleeing Boers, hacking and stabbing. Lance-Corporal Kelly of the 5th Lancers thrust through two Boers riding the same pony. On the second pass, one officer reported that he saw several incidents where Boers pleaded for their lives and generous Lancers turned aside their spear points, only to be fired on by those they had spared as they rode on. Eventually many of the Boers escaped into the broken country to the north. A little more cunning on the part of the British commander in blocking all exits and the whole Boer force would have been caught. As it was, Kock died on the battlefield and the Boers lost at a minimum 65 killed and 150 taken prisoner, among whom were 40 wounded. The Boer casualties may have been as high as 360. The British lost four officers and 46 men killed and 31 officers and 182 men wounded.

During the artillery bombardment, the twelve 15-pdr guns had fired 423 rounds and the six 7-pdr guns, 74 rounds. Over 61,000 rounds of rifle ammunition had been fired.

Strategically, the effect of the battle was nullified by the subsequent withdrawal of the British forces to Ladysmith; and the Boers soon reoccupied Elandslaagte. The immediate morale effect on the Boers was, however, significant. A few more decisive actions of this kind might have settled the war early on.

In a wargames representation, the Devons (seven companies), the Gordons (five companies) and the Manchesters (eight companies) can be represented by a total strength of 50 or so figures, assuming some depletion of the ranks due to sickness and other duties. Some accounts say only four companies of the Manchesters took part in the actual action. The cavalry, including 338 men in the five squadrons of the Imperial Light Horse, also need 50 or so figures. The artillery (20 officers and 532 men, including the Natal Volunteer Battery) have 22 figures. The Boer player has to be content with only 40 figures in all.

PAARDEBERG: A STRATEGIC VIEW

The actions around Paardeberg from 18–27 February 1900 provide a good illustration of how map movement can be used to mount a wargame. Three copies of the relevant maps are required, one for each side and one for the umpire. Both players mark their respective movements on their own maps and the umpire then determines which conflicts occur and what to tell each side. The British objective is to use its various forces to encircle the Boers while the Boers aim to break out with the assistance of a relieving force. The Boers have a central force of 4,000 men under Cronje and a secondary force of 3,500 men in various commandos nearby. The British have 20,000 men.

Cronje's wagon train (15mm scale).

Map movement is best done at brigade or commando level, with each movement block representing about 2,000 men. A map grid of ten-mile squares is appropriate. When the umpire has determined when contact has been achieved, the action can be transferred from the map to the wargames table. Suitable movement blocks for the manoeuvres near Paardeberg are commandos for Cronje (two), De Wet, De Beer and Steyn (Bethlehem commando); while the British have 13th Brigade (Major-General Knox), 14th Brigade (Major-General Chermside), 18th Brigade (Brigadier-General Stephenson), 19th Brigade (Brigadier-General

Cronje's Column heads for Paardeberg (15mm scale).

Smith-Dorien), the Highland Brigade (Brigadier-General Mac-Donald) and a weakened Cavalry Division under Lieutenant-General French. Mounted Infantry detachments are used as scouts. On good tracks, wagons, infantry and artillery move two squares a day; mounted troops move four squares; and couriers eight squares. Movement is halved across country or in adverse conditions.

By the relief of Kimberley, Roberts had opened the way for an advance on the Orange Free State. General Cronje realised this and, breaking up his camp near Magersfontein, began to retreat towards the state capital of Bloemfontein. Of all the lines of retreat he could have chosen, the route he adopted along the Modder River offered the least chance of success. It gave his slow column no appreciable start and separated him from other Boer forces. By 16 October the British forces were already practically athwart his path. Lieutenant-General Sir Kelly-Kenney's Sixth Division (13th and 18th brigades), which included a strong force of mounted infantry, contacted Cronje's column but, after being driven off by artillery fire at Klip Kraal, lost contact with the Boers for several hours. However, the leading elements of the Division caught up with the Boer rearguard at 10.00am on 17 February, north of Paardeberg Drift. Believing that the British could not reach Paardeberg Drift in force by nightfall, Cronje took his time to cross Vendutie Drift, allowing his men to stop for lunch and a nap. After passing Klip Drift, 200 Free Staters left Cronje's force, to be followed later by 500 Cape Colony volunteers.

Then came a dramatic development. French, who had ridden hard from Kimberley with the remnants of the Cavalry Division (1,500 men) suddenly appeared from the north, and the Boer advance was checked by artillery fire at 2,500 yards range from the Divisional guns. In the late afternoon Colonel O C Hannay and the Mounted Infantry arrived from the south and the south-east and quickly occupied the kopjes. The bulk of the army, pushed on by Kitchener, reached the banks of the Modder after marching for most of the night and deployed close to Cronje's overnight laager. Cronje could have escaped during the night if he had been prepared to abandon his wagons. He could have brushed the weak Cavalry Division aside and struck northwards to link up with Ferreira or he could have joined De Wet to the east. But he felt he had a good defensive position and nothing would convince him to leave. He had with him nearly 4,000 men and the prospect of reinforcements. His position was concentrated in an area of only one square mile but it was well entrenched. Tollie De Beer's commando took up position to the north on the high ground at Koedoesrand to protect the way ahead.

Before dawn on 18 February, the British troops moved up to

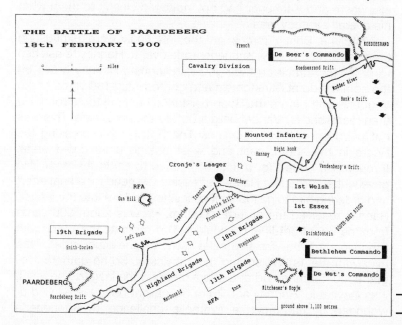

THE BATTLE OF PAARDEBERG
18th FEBRUARY 1900

strategic positions. Colvile's 9th Division (19th and Highland Brigades) had joined the 6th Division during the night. On the right flank, Hannay's Mounted Infantry (1,660 men) started for Koedoesrand Drift at 4.00am. His force had been weakened by the detachment of nearly a thousand men to support the western operations. A further 550 men were detached to secure a large kopje, later known as Kitchener's Kopje. At 5.30am Kitchener was delighted to see that Cronje was still in his laager. He realised that if he moved quickly he could surround him. At 6.30am the British artillery went into action to silence the Boer guns and by 7.00am the infantry were able to advance northeastwards along the river. Hannay's troops were able to seize the two small drifts called Vandenberg's and Bank's and link up with the 12th Lancers from French's Division on the north bank. Supported by the Welsh and the Essex regiments under Brigadier-General Stephenson, they worked their way downstream towards the laager. The pincer movement was closing fast.

However, between 8.30am and 10.30am, the Bethlehem commando at Stinkfontein caused some delay to the pincer assault, giving Cronje valuable time to improve his defences. On the left flank, the British suffered severe casualties in their open advance but nearly took the Boer positions. Individual commanders attacked piecemeal for they were unaware of the state of battle elsewhere and Kitchener had not indicated clearly to them what his overall strategy was.

While the British forces on the left flank were exhausting themselves in this way, Kitchener rode over to Hannay to ask him to press the attack on the right. In response, Hannay drove off the commando at Stinkfontein and got to within 100 yards of the Vendutie Drift before the Boer resistance became too strong. He then recrossed by Vandenberg's Drift to the north side. This was not good enough for Kitchener. The British were pressing the Boers in the south, east and west but no units could break through. He hoped for a final hard drive to settle the issue. He gave written orders to Hannay stressing the need for a final effort and ordering him to gallop up if necessary and fire into the laager. Hannay received the order at 3.00pm. He was about 500 yards from the Boer front-line. He could not co-ordinate the attack with Stephenson, who was two miles away on the other side of the river with the 1st Welsh and the 1st Essex. So he gathered together 40–50 men and at 3.30pm charged the laager. The result was inevitable. Hannay and 20 of his gallant band were killed or wounded. It had been a symbolic challenge to Kitchener's indifference to life (*Soldiers of the Queen*, March 1987, p. 22).

A further disruption of the attack came at 5.00pm when De Wet, in a brilliant tactical coup, stole up from the south-east with only 300 men and stormed Kitchener's Kopje. By 6.00pm all the ridges from the kopje to Koedoesrand were occupied by the Boers. This virtually concluded the battle but Kitchener would not cease hostilities until night fell. By then he had lost 24 officers and 279 men killed, 59 officers and 847 men wounded and two officers and 59 men missing. The Boers had lost 100 killed and 250 wounded.

De Wet actually held the kopje and the south-east ridge until Wednesday 21 February and allowed 100 Boers to escape from the laager. The rest elected to remain with Cronje.

After the events of the 18th – the fact that it was a Sunday had not prevented a fierce resistance by the Boers – the siege of the Boer laager settled into something of a stalemate. The Boer position was protected by a network of rifle pits extending for two miles below the laager and a mile above it. These were protected by stone ramparts with loopholes and connected with flanking trenches constructed from the numerous dongas that joined the river at right angles.

The British troops were forced to entrench at a considerable distance from the laager and to move forward very cautiously, digging fresh cover as best they could. The Royal Engineers then dug proper trenches behind the firing line, lying down as they worked.

The unwillingness of the British to fire on the women and children in the laager became less pronounced. Tactically, the superior artillery of the British (50 guns) enabled the laager to be shelled at will from Gun Hill and other positions, many horses and oxen being killed. The decaying carcasses of these animals were an early form of biological warfare: they were said to have caused many cases of enteric fever, to which the Boers had previously shown far more resistance than the British. The Boers thus had to suffer the torment of a bombardment by lyddite shells in a small area, without sanitation and with only the polluted water of the Modder River to drink.

However, the Boers were reassured by a heliogram from Commandant Froneman: "I am here with General De Wet and Commandant Cronje. Have good cheer. I am waiting for reinforcements. Tell the burghers to find courage in Psalm xxvii". But the reinforcements were beaten back. After 10 days the Boers had had enough. A disheartened Cronje and 4,000 Boers, including 60 women, surrendered to Roberts on 27 February – the anniversary of Majuba. For the first time the highly mobile Boers had been outmanoeuvred. The plan of encirclement had worked – but at the price of 1,250 British troops dead or wounded. Roberts is reputed to have told Kitchener in no uncertain terms that his job was to act as Chief of Staff. He did not expect him to fight the Commander-in-Chief's battles.

DUIVELSKLOOF (DEVIL'S CHASM): A SKIRMISH ACTION

By May 1901 the war had ceased to be an affair between gentlemen. Boer guerillas, loosely co-ordinated under Botha, De la Rey, De Wet and Smuts, had adopted tactics of tearing up railway lines, cutting telegraph wires, seizing convoys and over-running weakly-held posts. They were sustained by the support of the local civilian population. In retaliation the British began a scorched-earth policy in areas believed to have harboured the guerillas. Blockhouses were constructed within firing range of each other and linked by barbed wire. In constant sweeps across the veld, farms were burnt, livestock killed and women and children transported to concentration camps. This had two effects. First, a sense of bitterness consumed the Boers at large. Second, they were freed of the responsibility to defend their families and could thus remain in the field longer. Some accounts suggest that the scorched-earth policy was self-defeating in the sense that it may have prolonged the war by over a year.

The Boers were no less effective at guerilla action than they had been in formal battles in the earlier part of the war. Between October 1900 and September 1901 Boer guerillas tore up railway lines on average 16 times a month: in three months it was 30 times or more. But the great drives of British columns of mounted infantry, combined with the blockhouse system, made it more and more difficult for the Boers to evade capture. By mid-1901 the commandos were in dire straits: many horses had died and the men were in rags and, unless they could find a small British convoy to raid, lacked food and ammunition. It was a time of frustration on both sides, interspersed with short episodes of violent action. One such episode was the skirmish at Duivelskloof in the north-east Transvaal on the night of 5/6 August 1901.

Our scenario concerns a detachment from the Bushveldt Carbineers (BVC), a highly irregular mounted unit formed specific-ally to counteract the guerilla tactics of the Boers. The following men from B Squadron of the BVC have been assigned to a night raid on a Boer farm.

Name	No.	Status	Fire	Mêlée
Captain Percy Frederick Hunt		Veteran	8	8
Sergeant Frank Eland	148	Veteran	7	7
Corporal E G Browne	132	Veteran	7	6
Trooper G A Heath	241	Veteran	6	6
Trooper Albert van der Westhuizen	79	Veteran	6	6
Trooper F Yates	128	Average	6	5
Trooper J J van Blerk	259	Average	5	6
Shoesmith S H Haslett	72	Average	5	5
Trooper John S Silke	158	Average	5	5
Trumpeter H Hillsdon	349	Novice	4	4

The BVC are effective fighters but have an unconventional approach to discipline. Nearly half the men in the unit as a whole are Australians, nearly one-third are British and one-sixth South African, with a smattering of Americans, New Zealanders and other nationalities. The patrol is a fair sample of this cosmopolitan unit. Captain Hunt was formerly an officer in the 10th Hussars while almost all his men have had previous military experience in other units. Sergeant Eland is a brave and resourceful NCO. The only novice in the detail is Trumpeter Hillsdon. They wear the normal khaki outfit with the distinguishing mark of a dark green hat band.

Captain Hunt has received information from local Kaffirs that there are Boers at Viljoen's farm near Duivelskloof. He plans an early morning raid, while the Boers are asleep. Guided by a Kaffir, the BVC men dismount to the south of the farm and begin to creep towards their objective, which is seen in the early morning light to be a rough square of farm buildings surrounding a yard. On the north side is a barn alongside which is a pile of hay. The Kaffir points out the gully running near the main farm building and says that this curves steeply to the west behind the farm, thus providing excellent cover for any Boers fleeing from the farm.

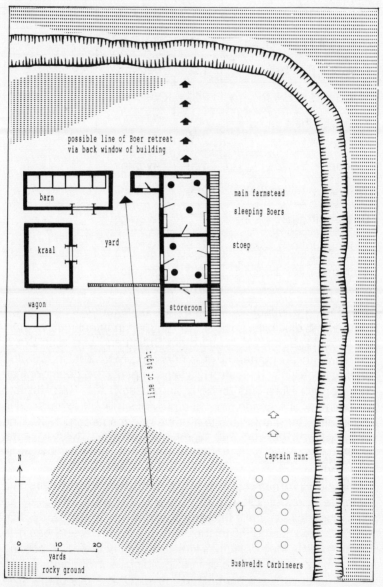

possible line of Boer retreat
via back window of building

barn

main farmstead

sleeping Boers

kraal

yard

stoep

wagon

storeroom

line of sight

Captain Hunt

N

0   10   20
yards
rocky ground

Bushveldt Carbineers

About 40 yards to the south of the farm is a solid shoulder jutting out of the slope, which provides a good firing position to cover the attack. A spruit runs 20 yards east of the farm.

Captain Hunt orders his men not to open fire until he has fired a signal shot from his pistol. He tells them to advance with extreme care: if the Boers are alerted the party could be caught in the open. Apart from Mauser rifles, the enemy may even have a Maxim gun with them, the product of a raid on a British outpost the week before. However, there is a problem. Trumpeter Hillsdon is extremely nervous, this being his first action, and his itchy trigger finger may spell disaster for the party. His safety catch is off. Captain Hunt can rely entirely on loyal support from Sergeant Eland but Corporal Browne has a grudge against all officers and may well hang back in action and encourage others to do the same.

Accompanying the BVC patrol are a group of Kaffirs, who will take to the woods as soon as the first shot is fired and who will remain there unless they see an opportunity to pick off any of the hated Boers who have executed several of their fellows for helping the British.

Meanwhile, the Boer guerillas in the farm are fast asleep, their Mauser rifles close beside them. They have no Maxim gun. Eight thin ponies are stabled in the yard. The guerillas are short on ammunition – only ten rounds apiece – and are hungry and ragged. They would love to get their hands on new equipment and on plump British horses. But above all, they do not intend to be caught by the British: they will, in all unfavourable circumstances, withdraw so that they can live to fight another day. They will resort to any tactics to resist capture, including ruses de guerre such as hoisting a white flag and then firing on the attackers as they advance to take the surrender. The group consists of:

Name	Status	Fire	Mêlée	Age
Veldkornet Barend Viljoen	Veteran	9	6	40
J J van der Merwe	Veteran	9	6	34
John Geyser	Veteran	9	4	72
Karl Smith	Veteran	8	6	21
Hendrik Visser	Veteran	7	4	26
Piet J Booysen	Novice	6	3	15

Viljoen is an exceptionally capable Boer officer, as brave as a lion and as cunning as a fox. Hennie Visser, by contrast, is extremely dim-witted. Piet Booysen is a young lad determined to avenge the death of his father and two brothers. The other Boers are typical Takhaars. Geyser is a Dopper and even more religious than his fellow burghers, all of whom have blind faith that God is on their side no matter what they may do.

As far as the mechanics of the game are concerned, Trumpeter Hillsdon has a 10% chance each move of firing his rifle prematurely and thus alerting the Boers. All weapons are assumed to be fully loaded. Each Boer may fire only 10 shots in the game, unless he can capture a British rifle and ammunition. Otherwise, normal skirmish rules are used. The longer the Carbineers have to creep round the building, the better their chance of cutting the Boers off but also the better the chance of Trumpeter Hillsdon firing prematurely.

Points to the Boer side in the game are two for every Khaki killed; one for every Khaki captured or seriously wounded; one for every rifle or horse captured from the Khakis; and nought for every Khaki who survives the action with a light wound or unscathed. British points are four for every Boer killed; two for every Boer captured, one for every Boer seriously wounded but escaping; and nought for every Boer escaping with a light wound or unwounded.

The points system does not encourage the taking of prisoners, but players are expected to play according to character. The British player may feel, for example, that a regular officer such as Captain Hunt would have to have a very good reason for shooting a prisoner, young or old. If a good reason – such as an abuse of the white flag – does occur in the skirmish one or two of the Khakis might of course serve up their own justice.

In the real action on which this scenario is based, the Boers were not caught napping. Captain Hunt was seriously wounded and Sergeant Eland killed. The fire from the Boers was so hot that the BVC survivors retreated. Yates was not among them. Most of the Boers escaped. When a BVC rescue party arrived, Viljoen and another Boer were found shot dead in the farmhouse. Several Africans were also found dead near the farm. Trooper Yates had been stripped and tied to a wagon while the body of Captain Hunt had been stripped and mutilated. It is not clear if the Boers or local blacks were responsible for mutilating the body, or that of Viljoen's, which had suffered a similar fate.

When a Boer, Hendrik Visser, was later captured wearing

Captain Hunt's uniform, he was shot on the spot by the Bush-veldt Carbineers without a trial. Lieutenants Harry Morant and Peter Handcock were subsequently found guilty of this and other military crimes by a court-martial and were executed at Pieter-maritzburg by a firing party from the Cameron Highlanders. Their real crime was that they did not have the rank to legitimise their actions: Kitchener hated to have indiscipline openly attributed to troops under his command.

SOUNDING THE "CHARGE."
A cavalry trumpeter carries both a bugle for field calls, and a trumpet for the more elaborate camp and barrack calls.

Sounding the 'Charge'.

The war had seen the deployment of 450,000 British and Dominion troops in South Africa, comprising 256,000 British regulars, 109,000 volunteers, 31,000 from the self-governing colonies and 53,000 recruited locally. Some 24,000 (5%) lives had been lost, 16,000 through disease. Over £220,000,000 had been added to the National Debt. Official casualty figures were 7,792 killed and 20,811 wounded, of whom 701 and 1,668 respectively were officers. Some 13,000 officers and men had died of enteric fever and other diseases. A further 64,000 men had been invalided home.

Some 87,000 burghers (60% from the Transvaal) had fought against the British, plus 2,700 regulars, 2,700 foreign volunteers and 13,000 rebels from the Cape and Natal. Between 4,000 and 7,000 republican fighters had died. In May 1902, there were 24,000 Boer prisoners of war in camps overseas and a further 7,000 in confinement or on parole in South Africa. The real loss, however, was the 28,000 Boer civilians, many of them children under 16, who perished in British concentration camps during the protracted guerilla phase of the war.

By the end of the war, 116,000 Africans had been removed to concentration camps, as the scorched-earth policy destroyed their livelihood. Over 14,000 black refugees died in the camps.

During the war, the British Army Remount Department provided approximately 520,000 horses and 150,000 mules, of which 350,000 horses and 50,000 mules were killed in action or died of disease or ill-use.

APPENDIX A
DIARY OF EVENTS OF THE SECOND
ANGLO-BOER WAR 1899–1902

BOER OFFENSIVES

October 1899

12th	Boers invade Natal and Cape Colony
14th	Boers begin siege of Kimberley (Lieutenant-Colonel Robert Kekewich) and Mafeking (Colonel R S S Baden-Powell)
20th	Battle of Talana Hill: Major-General Sir W Penn Symons mortally wounded; Lieutenant-Colonel B D Möller and 200 cavalry pursuing retreating Boers captured
21st	Battle of Elandslaagte
23rd	Battle of Dundee
24th	Battle of Rietfontein: inconclusive action
30th	'Mournful Monday': siege of Ladysmith begins after Commandant-General Piet Joubert outmanoeuvres Lieutenant-General Sir George White at Modderspruit; Lieutenant-Colonel F R C Carleton forced to surrender at Nicholson's Nek with 950 men
31st	General Sir Redvers Buller appointed Commander-in-Chief of British forces in South Africa

November 1899

15th	Armoured train incident near Chieveley: Winston Churchill taken prisoner
21st	Battle of Willow Grange: tactical victory for Boers
22nd	Battle of Mooi River: stand-off
	Lord Methuen commands field force in Orange River area; arrival of British reinforcements, including Australian and Canadian contingents;

three expensive British victories at:

23rd	Battle of Belmont
25th	Battle of Graspan (Enslin)
	Three Kgatla regiments attack Boer laager at Derdepoort in western Transvaal in the 'Second Battle of Blood River'
28th	Battle of Modder River (actually the Riet River)

December 1899

	'Black Week': three major British defeats at:
10th	Battle of the Stormberg (Lieutenant-General Sir William Gatacre)
11th	Battle of Magersfontein (Lieutenant-General Lord Methuen)
15th	Battle of Colenso (General Sir Redvers Buller)
18th	Field Marshal Lord Roberts appointed to succeed Buller, with General Lord Kitchener as Chief of Staff; army reserve called up
22nd	Boer reprisals for Derdepoort: 150 Kgatla killed

ROBERTS IN COMMAND

January 1900

6th	Boers attack Caesar's Camp and Wagon Hill (Platrand) at Ladysmith
24th	Battle of Spionkop: Major-General E R P Woodgate killed

February 1900

5th–7th	Vaalkrantz captured by British and then evacuated
15th	Major-General John French's cavalry division relieves Kimberley
18th	Battle of Paardeberg: uncoordinated British attacks defeated in detail
27th	Commandant-General Piet Cronje and 4,000 Boers surrender to Roberts at Paardeberg
28th	Buller relieves Ladysmith: the garrison of 13,500 of all ranks had suffered 600 casualties during the siege

March 1900

7th	Battle of Poplar Grove
10th	Battle of Dreifontein
13th	Roberts' columns occupy Bloemfontein
31st	General Christiaan De Wet ambushes force under Major-General R G Broadwood at Sannah's Post (Kornspruit): 400 prisoners taken by Boers

April 1900

4th	Surrender of 600 Royal Irish Rifles to De Wet at Reddersburg
9th-25th	Successful defence of Wepener: 1,900 men of Brabant's Horse besieged by De Wet

May 1900

17th	Relief of Mafeking
18th–31st	Naval Contingent with 4.7-inch and 12-pdr guns take part in Royal Military Tournament in London
24th	Annexation of Orange Free State: renamed Orange River Colony
31st	Pretoria occupied
	Surrender of 13th Battalion of Imperial Yeomanry to Piet De Wet at Lindley

June 1900

5th	Capture of Pretoria: 3,000 British prisoners released
11th–12th	Battle of Diamond Hill east of Pretoria

July 1900

11th	Surrender of 190 Scots Greys and Lincolns to De Wet at Zilikat's Nek
30th	Commandant-General Marthinus Prinsloo surrenders with 4,500 men at Brandwater Basin

Scattered fighting in spite of captured Boers signing oath; Roberts authorises burning of farms used by guerillas; Boer women, children and black

servants to be taken to internment camps from
September; start of first hunt for De Wet

August 1900

27th	Buller defeats General Louis Botha at Bergendal (Dalmanutha)
30th	Release of 2,000 British prisoners at Nooitgedacht

September 1900

1st	The Transvaal is annexed to the British Crown
11th	Kruger leaves for Europe
24th	British reach Mozambican border at Komatipoort

THE GUERILLA WAR

November 1900

23rd	De Wet captures 450 prisoners at Dewetsdorp
29th	British regard war as practically over: Kitchener succeeds Roberts as Commander-in-Chief in South Africa

December 1900

	Colonial Office asks New Zealand authorities not to include Maoris in volunteer force embarking for South Africa
13th	General Koos De la Rey and State Attorney Jan Smuts surprise Major-General R A P Clements at Nooitgedacht
16th	Commandos under General P H Kritzinger and Judge Barry Hertzog enter Cape Colony on anniversary of Battle of Blood River 1838
20th	Martial law declared in parts of the Cape
29th	Capture of Helvetia Post in east Transvaal by Commandant Ben Viljoen

January 1901

	New system of blockhouses instituted by Kitchener
4th–31st	Attacks by De la Rey, De Wet, Smuts and General

Christiaan Beyers

31st   Smuts captures Modderfontein: massacre of 100 Africans

February 1901

7th   An additional 30,000 mounted troops despatched to South Africa

10th–28th   De Wet's 'invasion' of Cape Colony

March 1901

Failure of peace negotiations at Middelburg

April 1901

10th   Major-General E L Elliot begins major drive in Orange River Colony

July 1901

6th   300 troopers of the 5th Victoria Rifles bush-whacked near Wilmansrust in Transvaal: Boer attackers wear khaki uniforms

18th   Drive northwards in Cape Colony begins

August 1901

5th   Skirmish at Duivelskloof (Devil's Chasm)

7th   Kitchener proclaims banishment for any Boer leaders caught in arms after 15th September 1901

12th   Kritzinger driven out of Cape Colony

September 1901

3rd   Smuts invades Cape Colony via Kiba Drift

5th   Commandant Johannes Lotter and 120 rebel Afrikaners from the Cape captured by Colonel H Scobell's column

17th   Smuts defeats 17th Lancers at Elands River Poort: victory due in part to use of khaki uniforms by Boers

Botha defeats force under Lieutenant-Colonel Hubert Gough at Blood River Poort

October 1901

9th   Martial law extended in the Cape
11th   Commandant Lotter executed for shooting loyal blacks and other acts; Commandant Gideon Scheepers captured

December 1901

7th   National Scout Corps inaugurated in Transvaal (hands-upper Boers fighting on the British side); equivalent in Free State known as Orange River Volunteers
25th   De Wet defeats 400 Yeomanry under Major Williams at Tweefontein

January 1902

17th   Scheepers executed at Graaf Reiner as a Cape rebel

February 1902

6th   New drive begins against De Wet in Orange River Colony
7th   De Wet breaks through blockhouse line
27th   General Lucas Meyer's commando of 650 men captured in laager at Lang Riet
             Lieutenants Harry Morant and Peter Handcock of Bushveldt Carbineers executed at Pietersburg

March 1902

7th   Lord Methuen captured by De Wet at Tweebosch; column of 1200 men with four guns virtually wiped out
24th   Drive begins in west Transvaal

April 1902

4th   Smuts invests Ookiep; town not relieved until May 3rd
11th   Battle of Rooiwal; Boer mounted charge repulsed

May 1902

6th — Zulu (Qulusi) attack on Boer force at Holkrantz near Vryheid; 56 of the 70 Boers killed; Boers seriously alarmed by this development

15th — Opening of Vereeniging conference

18th — Boer delegates, including Schalk Burger, acting President of the Transvaal, and Marthinus Steyn the Free State President, meet Kitchener and Sir Alfred Milner at Pretoria

31st — Final meeting at Vereeniging; Boer delegates sign surrender terms; British annexation accepted and authority of the British monarch, Edward VII, recognised; £3 million to be provided to the Boers by H M Treasury for economic reconstruction

May 1961

31st — Dr Verwoerd leads South Africa out of the British Commonwealth, 59 years to the day after the signing of the peace of Vereeniging

APPENDIX B

GLOSSARY

Afrikaner	originally used after the Great Trek for settlers of Dutch origin who remained in Cape Colony; now as 'Afrikaner' used to describe any descendant of Dutch settlers in South Africa
agterryer	an after-rider, black servant of a Boer
berg	mountain
biltong	thin strips of lean meat dried in the sun
bitter-ender	'bittereinder', a Boer fighter who remained in the field to the end of the war
Boer	originally applied to the settlers of Dutch origin who left the Cape in the Great Trek and set up the two republics in the interior; literal meaning is farmer, peasant, hence knave or jack (in a pack of cards)
brandwacht	'fire-guard' or Boer sentry
bywoner	poor white, landless Boer
burgher	a male citizen of the Boer republics with the rights and privileges of citizenship and a duty to undertake military service if called up
commandant	'kommandant' in Afrikaans, senior officer in charge of a commando
Commandant-General	army commander
commandeer	a process of obtaining supplies for a commando from local inhabitants; government vouchers were usually given in exchange
commando	a 'regiment' of burghers, usually from a particular district
donga	a dry river bed or gully
Doppers	Boer religious sect
dorp	village

drift	a ford across a river
fontein	spring or fountain
hands-upper	'hensopper' in Afrikaans, a surrendered burgher who helped the British
inspan	to yoke oxen to a wagon
Kaffir	from the Arabic for 'infidel'; strictly a particular tribe but became a term used by the Boers and the British for any black; now an extremely offensive description
klip	stone, pebble, rock or slang for diamond
kloof	chasm, gulf or ravine
kop	literally a 'head'; hence a prominent hill providing a good observation point (or a football terrace); 'koppie' or 'kopje' a small conical hill
kraal	African settlement or enclosure
krantz	rock ledge or crown of rocks on top of a mountain
krijgsraad	tactical meeting before action; council of war
kripvreter	a stall-fed horse, ie a Boer 'staff officer'
laager	Boer encampment of wagons formed in a circle
laagte	valley or glen, meadowland or shallow dip
mealies	maize, sweetcorn
nek	saddle connecting two hills
outspan	to unyoke oxen from wagons
pont	a ferry or ferry-boat, a pontoon
poort	'gate' in town names, or a mountain pass
rand	ridge, high land above a river valley
schanz	parapet or defensive stone wall
spruit	ditch, stream or small tributary
stad	town
stoep	verandah of a Boer farmstead
takhaar	a backvelder, country Boer
Transvaal	across the Yellow River, more accurately tawny or ashen-coloured river
trek	movement of people or journey; 'trekboer', a farmer seeking new land away from British influence

Uitlander	a foreigner or outlander; the name given by the Boers to non-Boer whites in the Johannesburg area
Vechtgeneraal	fighting general, also 'veggeneraal'
veld	field, open country; 'veldkornet', field-cornet
voorlooper	leader of a span of oxen
vrou	Boer housewife, queen (in a pack of cards)
ZARPs	Zuid-Afrikaansche Republiek Politie (Transvaal Police)

APPENDIX C

BOER COMMANDOS AT THE OUTBREAK OF WAR

Transvaal

Commando	Commandant	Maximum Call Up 16–60
Bethal	–	781
Bloemhof	J F de Beer	958
Carolina	D Joubert	506
Ermelo	T Smuts	862
Heidelberg	J D Weilbach	1,578
Johannesburg	J de Meillon	1,000
Lichtenburg	H C W Vermaas	1,005
Lydenburg	H P Steenkamp	834
Marico	J L D Botha	1,265
Middelburg	C E Faurie	2,081
Potchefstroom	M J Wolmarans	3,379
Pretoria	D J E Erasmus	3,746
Piet Retief	C L Engelbrecht	432
Rustenberg	H P Malan	2,536
Standerton	J A Muller	1,130
Swaziland	A M van Staden	332
Utrecht	L Viljoen	775
Vryheid	L S Meijer	944
Wakkerstroom	J A Joubert	1,254
Waterberg	P T Potgieter	732
Wolmarasnstadt	F J Potgieter	772
Zoutpansberg	D J L H du Preez	1,287
	Total	28,189

Orange Free State

Commando	Maximum Call Up 16–60
Bethlehem	1,635
Bethulie	641
Bloemfontein	2,427
Boshof	1,148
Caledon-rivier	898

Fauresmith		1,638
Harrismith		896
Heilbron		1,826
Hoopstadt		891
Jacobsdal		322
Kroonstadt		2,325
Ladybrand		2,044
Moroka		273
Philippolis		408
Rouxville		1,116
Vrede		966
Wepener		616
Winburg		2,234
	Total	22,304

APPENDIX D

SELECT BIBLIOGRAPHY

Amery L S, *The Times History of the War in South Africa* 1899–1902, seven volumes (1900–1907)

Barnes R M, *A History of the Regiments and Uniforms of the British Army* (1972)

Barthorp M, *The Anglo-Boer Wars* (1987)

Battles of the Nineteenth Century, volumes VI and VII (1901)

Churchill W S, *Frontiers and Wars* (1962)

Comaroff J L, *The Boer War Diary of Sol T Plaatje: an African at Mafeking* (1973)

Davey A, *Breaker Morant and the Bushveldt Carbineers* (Cape Town, 1987)

Denton K, *The Breaker* (1979) – a good novel but not historically accurate

Fordham D and Todd P, *Private Tucker's Boer War Diary* (1980)

Gordon L L, *British Battles and Medals* (1979)

Haythornthwaite P J, *The Boer War* (1987) – Uniforms Illustrated no 19

Hillegas H C, *With the Boer Forces* (1900) – a US correspondent who visited the Boer side

Intelligence Division War Office, *Military Notes on the Dutch Republics of South Africa* (1899)

James L, *The Savage Wars: British Campaigns in Africa 1870–1920* (1985)

Lee E, *To the Bitter End* (1985)

Maurice F, *History of the War in South Africa 1899–1902* (1906) – excellent maps

Pakenham T, *The Boer War* (1979)

Pemberton W B, *Battles of the Boer War* (1975)

Reitz D, *Commando: A Boer Journal of the Boer War* (1929)

Soldiers of the Queen, journal of the Victorian Military Society, various issues

Warwick P, *Black People and the South African War 1899–1902* (1983)

Whitehouse H, *Battle in Africa 1879–1914* (1987)

Wilson H W, *With the Flag to Pretoria* (1900) – most of the illustrations in this book are from this source (four volumes)

INDEX